Effective Python

编写好Python的90个有效方法

[美] 布雷特·斯拉特金（Brett Slatkin）著 （第2版 英文版）

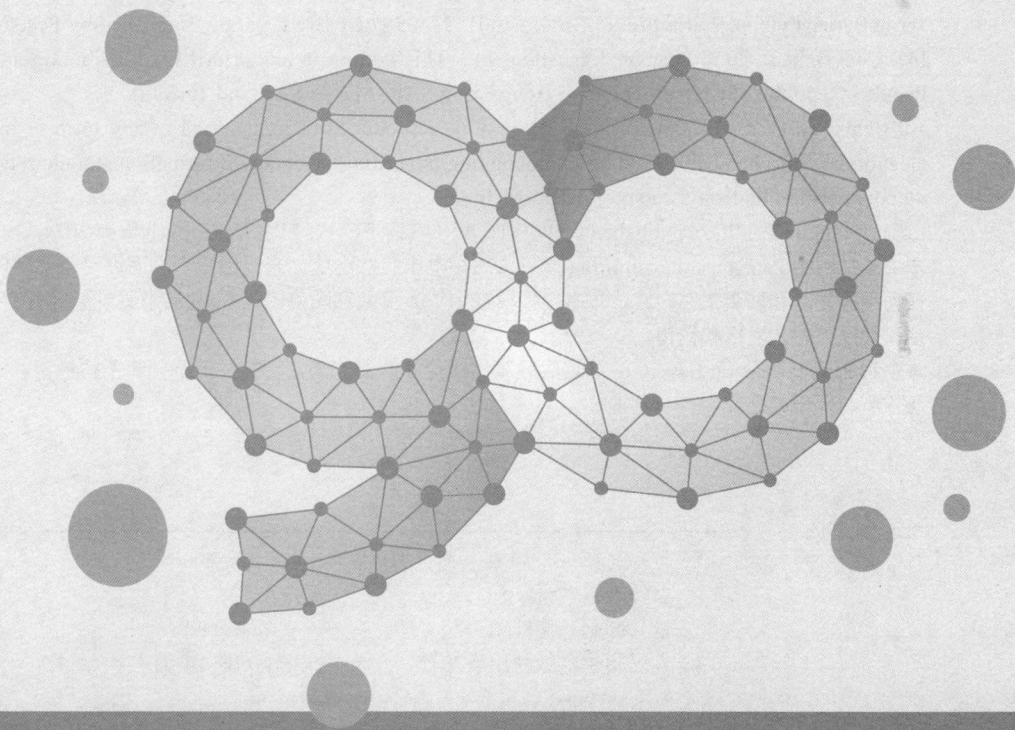

Effective Python:
90 Specific Ways to Write Better Python
(2nd Edition)

人民邮电出版社
北京

图书在版编目（CIP）数据

Effective Python：编写好Python的90个有效方法：第2版：英文 /（美）布雷特·斯拉特金（Brett Slatkin）著. -- 北京：人民邮电出版社，2024.3
    ISBN 978-7-115-63406-1

    Ⅰ. ①E… Ⅱ. ①布… Ⅲ. ①软件工具－程序设计－英文 Ⅳ. ①TP311.561

中国国家版本馆CIP数据核字(2024)第000603号

版权声明

◆ 著　　　［美］布雷特·斯拉特金（Brett Slatkin）
　　责任编辑　蒋 艳
　　责任印制　王 郁　胡 南
◆ 人民邮电出版社出版发行　　北京市丰台区成寿寺路 11 号
　　邮编　100164　电子邮件　315@ptpress.com.cn
　　网址　https://www.ptpress.com.cn
　　三河市兴达印务有限公司印刷
◆ 开本：787×1092　1/16
　　印张：28.25　　　　　　　2024 年 3 月第 1 版
　　字数：612 千字　　　　　2024 年 3 月河北第 1 次印刷
　　著作权合同登记号　图字：01-2023-4779 号

定价：128.00 元
读者服务热线：(010)81055410　印装质量热线：(010)81055316
反盗版热线：(010)81055315
广告经营许可证：京东市监广登字 20170147 号

内容提要

　　本书是 *Effective Python* 的第 2 版，基于 Python 3 进行了全面升级。本书分为 10 章，包含 90 个条目，形式简洁、表述准确。每个条目都基于对 Python 的独到见解，告诉读者如何高效地编写 Python 程序。与第 1 版相比，第 2 版只关注 Python 3，而不再兼顾 Python 2。根据 Python 语言引入的新特性，以及 Python 开发者所形成的新经验，对第 1 版中的某些思路与解决方案进行了修订，以更好地发挥 Python 的优势。本书删除了过时的知识点，并添加了对 Python 新特性的一些介绍。新版中增加了 31 个条目，并专门设立了几章来强调列表和字典、推导和生成器、元类和属性、稳定性和性能，以及测试和调试等重要主题。

　　建议读者在阅读本书之前应对 Python 语言有初步的了解。对于有一定经验的开发者以及对 Python 编程感兴趣的读者，本书可以帮助其更深入地理解 Python 语言，以成为更卓越、高效的 Python 开发者。

作者简介

布雷特·斯拉特金（Brett Slatkin）是 Google 首席软件工程师，Google Surveys 的联合技术创始人，也是 PubSubHubbub 协议的共同创造者之一。

此外，Slatkin 还发布了 Google 的第一个云计算产品——App Engine。早在 2006 年，Slatkin 就开始使用 Python 来管理 Google 庞大的服务器群。他拥有纽约哥伦比亚大学计算机工程学士学位。

本书赞誉

"自 2015 年第 1 版出版以来，我一直诚挚地向他人推荐这本书。本书第 2 版更新并补充了 Python 3 的相关内容，里面写满了各种实用的 Python 编程技巧，可以让具有不同开发水平的程序员受益。"

—— Wes McKinney，Python Pandas 项目创始人，Ursa Labs 主管

"如果你是从其他编程语言转到 Python 的程序员，那么你可以把这本书看作权威指南，它将教你怎样充分利用 Python 的各种特性。我已经使用 Python 将近 20 年了，仍然可以从这本书里学到很多有用的技巧，特别是与 Python 3 的新特性有关的技巧。本书中的方法都很实用，可以帮助我们判断什么样的代码才是 Python 风格的代码。"

—— Simon Willison，Django 联合创始人

"我已经用 Python 编程很多年，自认为对这门语言已经非常了解了，但是看过这本书中的各项技巧之后，我才发现自己写的代码仍然有很多地方可以改进。比如可以通过二分法在有序列表中搜索，让程序跑得更快；可以采用只能通过关键字形式来指定的参数把代码写得更加清晰、易读；可以通过星号表达式来拆分序列，减少出错率；可以通过 zip 并行迭代多个列表，使代码更具 Python 风格。第 2 版还有个优点，就是可以让我快速掌握 Python 3 的新特性，如 walrus 操作符、f-string 和 typing 模块等。"

—— Pamela Fox，Khan Academy 编程课创始人

"现在 Python 3 终于成为 Python 的标准版本，它其实已经经历了 8 次小的修订，并引入了许多新特性。Brett Slatkin 这次带来的 *Effective Python* 第 2 版，包含了大量新的 Python 习惯用法和很多条明确的建议，这些内容考虑到了从 Python 3.0 到 Python 3.8 之间的各个版本，现在可以让 Python 2 停一停了。本书前几章介绍了 Python 3 的许多新语法和新概念，如 strings 对象、byte 对象、f-string、赋值表达式（assignment expression，作者会提到关于这种表达式的几个你可能不知道的特殊名称），以及如何把 tuple 中的其他元素全都捕获下来。后面几章谈的主题比较大，其中有些内容我原来没听说过，还有一些我虽然听说过，但总找不到特别好的办法给别人讲解。比如第 6 章，其中有一个条目很棒，那就是条目 51。另外，作者还介绍了一个奇妙的方法，叫作__init_subclass__()，我原来不太熟悉。在介绍并发和并行的章节中，条目 53 讲得很好，当然作者还讲了

异步 I/O（asyncio）与协程（coroutine）的问题，而且讲得很对。另外就是谈稳定性和性能的那一章，其中条目 70 值得注意。总之，每一部分都写得很棒，作者把这些实践技巧讲解得相当清晰，以后我打算引用书里的一些内容，因为这些建议确实很有道理。如果今年只读一本 Python 书的话，那我肯定会选这一本。"

—— Mike Bayer，SQLAlchemy 创始人

"这是一本新手和有经验的程序员都适合阅读的好书。书中的范例代码写得很周详，作者解释得也相当精准、透彻。第 2 版更新了与 Python 3 有关的建议，这实在是太好了。我已经使用 Python 近 20 年了，但还是每看几页就能学到新的知识。本书中提供的建议对任何用 Python 编程的人都很有帮助。"

—— Titus Brown，加利福尼亚大学戴维斯分校副教授

"与第 1 版一样，Brett Slatkin 依然很好地将整个 Python 开发领域的各种经验，全部汇集到一本书中。他既谈了元类和并发等复杂的主题，也没有忘记稳定性、测试和协作开发等重要的基础知识。本书新版用大家都能接受的观点，讲述了什么叫作 Python 式的编程风格。"

—— Brandon Rhodes，python-patterns.guide 作者

前言

    Python 是一门强大的编程语言，它很有魅力，同时也很独特，所以理解起来可能有点儿困难。许多熟悉其他编程语言的程序员常常以有限的思维方式来对待 Python，而不是充分发挥 Python 的特性。还有一些程序员则过度使用 Python 的特性，这样写出来的程序以后可能会引发严重的问题。

    本书详细地介绍了如何采用符合 Python 风格的方式（Pythonic 方式）编写程序，这是使用 Python 的最佳方式。本书假设你已经对这门语言有了初步了解。新手程序员可以通过这本书学习 Python 功能的最佳实践，有经验的程序员则能够学会如何自信地运用 Python 中的新工具。

    本书的目标是让你能用 Python 开发出优秀的软件。

本书涵盖的内容

    本书的每一章都包含许多相互关联的条目。你可以按照自己的兴趣随意阅读不同的条目。每个条目都包含简明而具体的指导，解释了如何更有效地编写 Python 程序。条目里包括了关于要做什么、要避免什么、如何在各种做法之间取得平衡，以及为什么这是最佳选择的建议。条目之间相互引用，以帮助你在阅读过程中全面地了解这些建议所涉及的知识。

    本书第 2 版专注于 Python 3（参见条目 1），包括从 3.0 到 3.8 的各个版本。本书第 1 版中的大多数条目仍然收录在第 2 版中，并且做了修订，其中有些条目改动比较大。随着 Python 语言越来越成熟，最佳实践也发生了变化，因此两个版本中对某些问题的建议可能完全不同。Python 2 在 2020 年 1 月 1 日已经停止维护，如果你仍然主要使用 Python 2，那么第 1 版中的建议可能对你更有用。

    Python 采用了"自带电池"（batteries included）的方式来构建其标准库，与许多其他编程语言相比，它内置了大量常用的软件包，而不需要你到其他地方寻找重要的功能。许多 Python 内置软件包与 Python 的习惯用法密切相关，几乎可以将其看作是语言规范的一部分。Python 的标准模块太多，无法在本书中全部覆盖，但本书已经覆盖了其中那些需要了解和使用且用法比较关键的模块。

第 1 章：Pythonic 思维

Python 开发者社区用"Pythonic"这个形容词来描述具有某种特定风格的代码。这种风格是大家在使用 Python 语言编程并相互协作的过程中逐渐形成的。本章介绍如何采用这样的风格编写常见的 Python 代码。

第 2 章：列表和字典

在 Python 中，组织信息的最常见方式是将值序列存储在列表（list）中。与列表互补的结构是字典，它存储查找键与对应值的映射关系。本章介绍如何使用这些数据结构来编写程序。

第 3 章：函数

Python 中的函数具有多种特性，可以使程序员的工作更加轻松。Python 函数的某些特性与其他编程语言中函数的特性类似，但也有一些是 Python 特有的。本章介绍如何使用函数来表达开发者的意图，如何让代码更易复用，以及如何减少错误。

第 4 章：推导和生成器

Python 有一种特殊的语法，可以快速迭代列表、字典和集合，并据此生成相应的数据结构，这让我们能够在函数返回的这种结构上逐个访问根据原结构所派生出来的一系列值。本章介绍如何利用这种机制来提升程序效率并降低内存用量，同时提高代码的可读性。

第 5 章：类和接口

Python 是面向对象的语言。用 Python 编程时，经常需要编写新的类，并定义这些类应该如何通过其接口和继承关系与其他代码进行交互。本章介绍如何使用类来表达对象所应具备的行为。

第 6 章：元类和属性

元类和属性都是很强大的 Python 特性，然而，它们也有可能会让程序出现奇怪的行为与意外的效果。本章讲解这些机制的习惯用法，确保读者写出来的代码遵循"最少惊讶原则"（rule of least surprise）。

第 7 章：并发和并行

用 Python 编写并发程序很容易，这种程序可以同时执行许多不同的任务。Python 还可以通过系统调用、子进程和 C 语言扩展来实现并行处理。本章介绍这些 Python 特性应该在什么情况下使用。

第 8 章：稳定性和性能

Python 具有内置的功能和模块，可以让程序变得更加可靠。Python 还提供了一些工具，可以帮助你轻松提升程序的性能。本章介绍如何用 Python 优化程序，以最大程度地提高这些程序在正式运行环境中的可靠性和效率。

第 9 章：测试和调试

无论使用哪种编程语言，都应该对写出来的代码进行测试。然而，Python 的动态机制可能会增大程序在运行时出现错误的风险。幸运的是，Python 也让我们可以比较容易地编写测试代码和故障诊断程序。本章介绍 Python 中用于测试和调试的内置工具。

第 10 章：协作开发

协作开发 Python 程序需要谨慎考虑代码的编写方式。即使你只是独自开发，也需要了解如何使用其他人编写的模块。本章介绍协作开发 Python 程序的标准工具和最佳实践。

本书使用的约定

本书使用等宽字体表示 Python 代码。如果某一句代码比较长，则使用➥符号表示换行。代码中的某些部分如果跟当前要讲的主题关系不大，笔者会将这些代码删掉，并用...表示。你需要访问"获取源代码及勘误表"部分提到的网址以获取完整的范例代码，这样就能在自己的计算机上正确地运行这些例子了。

为了使范例代码更符合图书格式，且能够突出重要内容，笔者在 Python 风格指南的基础上做了一些修改。为了缩减范例代码篇幅，笔者还省略了嵌入式文档。你在开发自己的项目时不应该这样做，而是应该遵循 Python 风格指南（参见条目 2），并在源代码中撰写开发文档（参见条目 84）。

本书中的大部分代码都会有运行代码之后的相应输出。这里说的"输出"，指的是在交互式解释器中运行 Python 程序时，控制台或终端输出的内容。输出部分也使用等宽字体，并在上面有一行>>>符号（Python 解释器的提示符）。书中使用这个符号是想告诉你，把>>>上方的范例代码输入 Python shell（Python 解释器的界面）之后，会产生与>>>下方一样的输出信息。

此外，本书中还有一些文字上方虽然没有>>>符号，但也使用等宽字体。这些内容是指 Python 解释器之外的输出信息。这些信息的上方通常有$字符，表示笔者正在从类似 Bash 的命令行界面中运行程序。如果你是在 Windows 或其他类型的操作系统中运行这些程序的话，那么可能需要相应地调整程序名称和参数。

获取源代码和勘误表

将本书中的范例代码视为完整的程序并运行一遍是非常有用的。这可以让你有机会自己调整代码，并试着理解程序按照书中描述的方式运行的原因。所有的源代码都可以从本书网站（https://effectivepython.com）上下载。该网站还包括本书的勘误，以及提交勘误的方法。

致谢

在生活中，有许多人给了我指导、支持和鼓励，没有他们，就不会有这本书。

感谢 Effective Software Development 系列的顾问 Scott Meyers。我在 15 岁时初次阅读了 *Effective C++*，并爱上了编程。毫无疑问，我后来的求学经历以及第一份工作都得益于 Scott 的那本书。这次有机会编写 *Effective Python*，我感到很荣幸。

感谢本书第 2 版的技术审校人员 Andy Chu、Nick Cohron、Andrew Dolan、Asher Mancinelli 和 Alex Martelli，他们提供了深刻而详尽的反馈意见。感谢 Google 的同事审读这本书并给出建议。有了大家的帮助，这本书才更加通俗易懂。

感谢 Pearson 参与制作本书第 2 版的每一位工作人员。感谢执行编辑 Debra Williams 在整个过程中提供的支持。感谢各位团队成员的大力协助，他们是：开发编辑 Chris Zahn、营销经理 Stephane Nakib、文字编辑 Catherine Wilson、资深项目编辑 Lori Lyons，以及封面设计师 Chuti Prasertsith。

感谢支持我完成本书第 1 版的朋友：Trina MacDonald、Brett Cannon、Tavis Rudd、Mike Taylor、Leah Culver、Adrian Holovaty、Michael Levine、Marzia Niccolai、Ade Oshineye、Katrina Sostek、Tom Cirtin、Chris Zahn、Olivia Basegio、Stephane Nakib、Stephanie Geels、Julie Nahil 和 Toshiaki Kurokawa。感谢各位读者指出错误并提供修改建议。感谢各位译者将本书翻译成其他语言。

感谢与我交流并一起工作过的优秀的 Python 程序员：Anthony Baxter、Brett Cannon、Wesley Chun、Jeremy Hylton、Alex Martelli、Neal Norwitz、Guido van Rossum、Andy Smith、Greg Stein 和 Ka-Ping Yee。感谢你们引导我学习 Python。Python 拥有一个出色的开发者社区，能成为其中的一员，我感到很幸运。

感谢各位伙伴这些年来对我的关照。感谢 Kevin Gibbs 帮助我应对风险。感谢 Ken Ashcraft、Ryan Barrett 和 Jon McAlister 教我如何工作。感谢 Brad Fitzpatrick 帮助我提升工作能力。感谢 Paul McDonald 陪我一起创建我们的搞怪项目。感谢 Jeremy Ginsberg、Jack Hebert、John Skidgel、Evan Martin、Tony Chang、Troy Trimble、Tessa Pupius 和 Dylan Lorimer 帮助我学习。感谢 Sagnik Nandy 和 Waleed Ojeil 提供的指导。

感谢激发与培养我对代码与软件工程兴趣的各位老师：Ben Chelf、Glenn Cowan、Vince Hugo、Russ Lewin、Jon Stemmle、Derek Thomson 和 Daniel Wang。有了你们的指引，我才会努力提高编程技术，让自己有能力去教导他人。

感谢母亲帮助我找到了人生目标并鼓励我成为程序员。感谢我的兄弟、祖父母、各位亲戚以及儿时的朋友，从小你们就是我的榜样，让我体会到了成长的快乐。

最后，感谢我的妻子 Colleen，感谢她在人生旅程中给我的关爱、支持和欢笑。

目录

Chapter 1

Pythonic Thinking

The idioms of a programming language are defined by its users. Over the years, the Python community has come to use the adjective *Pythonic* to describe code that follows a particular style. The Pythonic style isn't regimented or enforced by the compiler. It has emerged over time through experience using the language and working with others. Python programmers prefer to be explicit, to choose simple over complex, and to maximize readability. (Type `import this` into your interpreter to read *The Zen of Python*.)

Programmers familiar with other languages may try to write Python as if it's C++, Java, or whatever they know best. New programmers may still be getting comfortable with the vast range of concepts that can be expressed in Python. It's important for you to know the best—the *Pythonic*—way to do the most common things in Python. These patterns will affect every program you write.

Item 1: Know Which Version of Python You're Using

Throughout this book, the majority of example code is in the syntax of Python 3.7 (released in June 2018). This book also provides some examples in the syntax of Python 3.8 (released in October 2019) to highlight new features that will be more widely available soon. This book does not cover Python 2.

Many computers come with multiple versions of the standard CPython runtime preinstalled. However, the default meaning of `python` on the command line may not be clear. `python` is usually an alias for `python2.7`, but it can sometimes be an alias for even older versions, like `python2.6` or `python2.5`. To find out exactly which version of Python you're using, you can use the `--version` flag:

```
$ python --version
Python 2.7.10
```

Python 3 is usually available under the name python3:

```
$ python3 --version
Python 3.8.0
```

You can also figure out the version of Python you're using at runtime by inspecting values in the sys built-in module:

```
import sys
print(sys.version_info)
print(sys.version)
```

```
>>>
sys.version_info(major=3, minor=8, micro=0,
➥releaselevel='final', serial=0)
3.8.0 (default, Oct 21 2019, 12:51:32)
[Clang 6.0 (clang-600.0.57)]
```

Python 3 is actively maintained by the Python core developers and community, and it is constantly being improved. Python 3 includes a variety of powerful new features that are covered in this book. The majority of Python's most common open source libraries are compatible with and focused on Python 3. I strongly encourage you to use Python 3 for all your Python projects.

Python 2 is scheduled for *end of life* after January 1, 2020, at which point all forms of bug fixes, security patches, and backports of features will cease. Using Python 2 after that date is a liability because it will no longer be officially maintained. If you're still stuck working in a Python 2 codebase, you should consider using helpful tools like 2to3 (preinstalled with Python) and six (available as a community package; see Item 82: "Know Where to Find Community-Built Modules") to help you make the transition to Python 3.

Things to Remember

✦ Python 3 is the most up-to-date and well-supported version of Python, and you should use it for your projects.

✦ Be sure that the command-line executable for running Python on your system is the version you expect it to be.

✦ Avoid Python 2 because it will no longer be maintained after January 1, 2020.

Item 2: Follow the PEP 8 Style Guide

Python Enhancement Proposal #8, otherwise known as PEP 8, is the style guide for how to format Python code. You are welcome to write Python code any way you want, as long as it has valid syntax. However, using a consistent style makes your code more approachable and easier to read. Sharing a common style with other Python programmers in the larger community facilitates collaboration on projects. But even if you are the only one who will ever read your code, following the style guide will make it easier for you to change things later, and can help you avoid many common errors.

PEP 8 provides a wealth of details about how to write clear Python code. It continues to be updated as the Python language evolves. It's worth reading the whole guide online Here are a few rules you should be sure to follow.

Whitespace

In Python, whitespace is syntactically significant. Python programmers are especially sensitive to the effects of whitespace on code clarity. Follow these guidelines related to whitespace:

- Use spaces instead of tabs for indentation.
- Use four spaces for each level of syntactically significant indenting.
- Lines should be 79 characters in length or less.
- Continuations of long expressions onto additional lines should be indented by four extra spaces from their normal indentation level.
- In a file, functions and classes should be separated by two blank lines.
- In a class, methods should be separated by one blank line.
- In a dictionary, put no whitespace between each key and colon, and put a single space before the corresponding value if it fits on the same line.
- Put one—and only one—space before and after the = operator in a variable assignment.
- For type annotations, ensure that there is no separation between the variable name and the colon, and use a space before the type information.

Naming

PEP 8 suggests unique styles of naming for different parts in the language. These conventions make it easy to distinguish which type corresponds to each name when reading code. Follow these guidelines related to naming:

- Functions, variables, and attributes should be in `lowercase_underscore` format.

- Protected instance attributes should be in `_leading_underscore` format.

- Private instance attributes should be in `__double_leading_underscore` format.

- Classes (including exceptions) should be in `CapitalizedWord` format.

- Module-level constants should be in `ALL_CAPS` format.

- Instance methods in classes should use `self`, which refers to the object, as the name of the first parameter.

- Class methods should use `cls`, which refers to the class, as the name of the first parameter.

Expressions and Statements

The Zen of Python states: "There should be one—and preferably only one—obvious way to do it." PEP 8 attempts to codify this style in its guidance for expressions and statements:

- Use inline negation (`if a is not b`) instead of negation of positive expressions (`if not a is b`).

- Don't check for empty containers or sequences (like `[]` or `''`) by comparing the length to zero (`if len(somelist) == 0`). Use `if not somelist` and assume that empty values will implicitly evaluate to `False`.

- The same thing goes for non-empty containers or sequences (like `[1]` or `'hi'`). The statement `if somelist` is implicitly `True` for non-empty values.

- Avoid single-line `if` statements, `for` and `while` loops, and `except` compound statements. Spread these over multiple lines for clarity.

- If you can't fit an expression on one line, surround it with parentheses and add line breaks and indentation to make it easier to read.

- Prefer surrounding multiline expressions with parentheses over using the \ line continuation character.

Imports

PEP 8 suggests some guidelines for how to import modules and use them in your code:

- Always put `import` statements (including `from x import y`) at the top of a file.

- Always use absolute names for modules when importing them, not names relative to the current module's own path. For example, to import the foo module from within the bar package, you should use `from bar import foo`, not just `import foo`.

- If you must do relative imports, use the explicit syntax `from . import foo`.

- Imports should be in sections in the following order: standard library modules, third-party modules, your own modules. Each subsection should have imports in alphabetical order.

Note

The Pylint tool is a popular static analyzer for Python source code. Pylint provides automated enforcement of the PEP 8 style guide and detects many other types of common errors in Python programs. Many IDEs and editors also include linting tools or support similar plug-ins.

Things to Remember

✦ Always follow the Python Enhancement Proposal #8 (PEP 8) style guide when writing Python code.

✦ Sharing a common style with the larger Python community facilitates collaboration with others.

✦ Using a consistent style makes it easier to modify your own code later.

Item 3: Know the Differences Between bytes and str

In Python, there are two types that represent sequences of character data: bytes and str. Instances of bytes contain raw, unsigned 8-bit values (often displayed in the ASCII encoding):

```
a = b'h\x65llo'
print(list(a))
print(a)

>>>
[104, 101, 108, 108, 111]
b'hello'
```

Instances of str contain Unicode *code points* that represent textual characters from human languages:

```
a = 'a\u0300 propos'
print(list(a))
print(a)

>>>
['a', '`', ' ', 'p', 'r', 'o', 'p', 'o', 's']
à propos
```

Importantly, str instances do not have an associated binary encoding, and bytes instances do not have an associated text encoding. To convert Unicode data to binary data, you must call the encode method of str. To convert binary data to Unicode data, you must call the decode method of bytes. You can explicitly specify the encoding you want to use for these methods, or accept the system default, which is commonly *UTF-8* (but not always—see more on that below).

When you're writing Python programs, it's important to do encoding and decoding of Unicode data at the furthest boundary of your interfaces; this approach is often called the *Unicode sandwich*. The core of your program should use the str type containing Unicode data and should not assume anything about character encodings. This approach allows you to be very accepting of alternative text encodings (such as *Latin-1*, *Shift JIS*, and *Big5*) while being strict about your output text encoding (ideally, UTF-8).

The split between character types leads to two common situations in Python code:

- You want to operate on raw 8-bit sequences that contain UTF-8-encoded strings (or some other encoding).
- You want to operate on Unicode strings that have no specific encoding.

You'll often need two helper functions to convert between these cases and to ensure that the type of input values matches your code's expectations.

The first function takes a bytes or str instance and always returns a str:

```
def to_str(bytes_or_str):
    if isinstance(bytes_or_str, bytes):
        value = bytes_or_str.decode('utf-8')
    else:
        value = bytes_or_str
    return value  # Instance of str
```

```
print(repr(to_str(b'foo')))
print(repr(to_str('bar')))
```

```
>>>
'foo'
'bar'
```

The second function takes a bytes or str instance and always returns a bytes:

```
def to_bytes(bytes_or_str):
    if isinstance(bytes_or_str, str):
        value = bytes_or_str.encode('utf-8')
    else:
        value = bytes_or_str
    return value  # Instance of bytes
```

```
print(repr(to_bytes(b'foo')))
print(repr(to_bytes('bar')))
```

There are two big gotchas when dealing with raw 8-bit values and Unicode strings in Python.

The first issue is that bytes and str seem to work the same way, but their instances are not compatible with each other, so you must be deliberate about the types of character sequences that you're passing around.

By using the + operator, you can add bytes to bytes and str to str, respectively:

```
print(b'one' + b'two')
print('one' + 'two')
```

```
>>>
b'onetwo'
onetwo
```

But you can't add str instances to bytes instances:

```
b'one' + 'two'
```

```
>>>
Traceback ...
TypeError: can't concat str to bytes
```

Nor can you add bytes instances to str instances:

```
'one' + b'two'
```

```
>>>
Traceback ...
TypeError: can only concatenate str (not "bytes") to str
```

By using binary operators, you can compare bytes to bytes and str to str, respectively:

```
assert b'red' > b'blue'
assert 'red' > 'blue'
```

But you can't compare a str instance to a bytes instance:

```
assert 'red' > b'blue'
```

```
>>>
Traceback ...
TypeError: '>' not supported between instances of 'str' and
➥'bytes'
```

Nor can you compare a bytes instance to a str instance:

```
assert b'blue' < 'red'
```

```
>>>
Traceback ...
TypeError: '<' not supported between instances of 'bytes'
➥and 'str'
```

Comparing bytes and str instances for equality will always evaluate to False, even when they contain exactly the same characters (in this case, ASCII-encoded "foo"):

```
print(b'foo' == 'foo')
```

```
>>>
False
```

The % operator works with format strings for each type, respectively:

```
print(b'red %s' % b'blue')
print('red %s' % 'blue')
```

```
>>>
b'red blue'
red blue
```

But you can't pass a str instance to a bytes format string because Python doesn't know what binary text encoding to use:

```
print(b'red %s' % 'blue')
```

```
>>>
Traceback ...
TypeError: %b requires a bytes-like object, or an object that
➥implements __bytes__, not 'str'
```

You *can* pass a bytes instance to a str format string using the % operator, but it doesn't do what you'd expect:

```
print('red %s' % b'blue')
```

```
>>>
red b'blue'
```

This code actually invokes the __repr__ method (see Item 75: "Use repr Strings for Debugging Output") on the bytes instance and substitutes that in place of the %s, which is why b'blue' remains escaped in the output.

The second issue is that operations involving file handles (returned by the open built-in function) default to requiring Unicode strings instead of raw bytes. This can cause surprising failures, especially for programmers accustomed to Python 2. For example, say that I want to write some binary data to a file. This seemingly simple code breaks:

```
with open('data.bin', 'w') as f:
    f.write(b'\xf1\xf2\xf3\xf4\xf5')
```

```
>>>
Traceback ...
TypeError: write() argument must be str, not bytes
```

The cause of the exception is that the file was opened in write text mode ('w') instead of write binary mode ('wb'). When a file is in text mode, write operations expect str instances containing Unicode data instead of bytes instances containing binary data. Here, I fix this by changing the open mode to 'wb':

```
with open('data.bin', 'wb') as f:
    f.write(b'\xf1\xf2\xf3\xf4\xf5')
```

A similar problem also exists for reading data from files. For example, here I try to read the binary file that was written above:

```
with open('data.bin', 'r') as f:
    data = f.read()
```

```
>>>
Traceback ...
UnicodeDecodeError: 'utf-8' codec can't decode byte 0xf1 in
➥position 0: invalid continuation byte
```

This fails because the file was opened in read text mode ('r') instead of read binary mode ('rb'). When a handle is in text mode, it uses the system's default text encoding to interpret binary data using the bytes.encode (for writing) and str.decode (for reading) methods. On most systems, the default encoding is UTF-8, which can't accept the binary data b'\xf1\xf2\xf3\xf4\xf5', thus causing the error above. Here, I solve this problem by changing the open mode to 'rb':

```
with open('data.bin', 'rb') as f:
    data = f.read()

assert data == b'\xf1\xf2\xf3\xf4\xf5'
```

Alternatively, I can explicitly specify the encoding parameter to the open function to make sure that I'm not surprised by any platform-specific behavior. For example, here I assume that the binary data in the file was actually meant to be a string encoded as 'cp1252' (a legacy Windows encoding):

```
with open('data.bin', 'r', encoding='cp1252') as f:
    data = f.read()

assert data == 'ñòóôõ'
```

The exception is gone, and the string interpretation of the file's contents is very different from what was returned when reading raw bytes. The lesson here is that you should check the default encoding on your system (using python3 -c 'import locale; print(locale.getpreferredencoding())') to understand how it differs from your expectations. When in doubt, you should explicitly pass the encoding parameter to open.

Things to Remember

+ bytes contains sequences of 8-bit values, and str contains sequences of Unicode code points.

+ Use helper functions to ensure that the inputs you operate on are the type of character sequence that you expect (8-bit values, UTF-8-encoded strings, Unicode code points, etc).

+ bytes and str instances can't be used together with operators (like >, ==, +, and %).

+ If you want to read or write binary data to/from a file, always open the file using a binary mode (like 'rb' or 'wb').

+ If you want to read or write Unicode data to/from a file, be careful about your system's default text encoding. Explicitly pass the encoding parameter to open if you want to avoid surprises.

Item 4: Prefer Interpolated F-Strings Over C-style Format Strings and str.format

Strings are present throughout Python codebases. They're used for rendering messages in user interfaces and command-line utilities. They're used for writing data to files and sockets. They're used for specifying what's gone wrong in Exception details (see Item 27: "Use Comprehensions Instead of map and filter"). They're used in debugging (see Item 80: "Consider Interactive Debugging with pdb" and Item 75: "Use repr Strings for Debugging Output").

Formatting is the process of combining predefined text with data values into a single human-readable message that's stored as a string. Python has four different ways of formatting strings that are built into the language and standard library. All but one of them, which is covered last in this item, have serious shortcomings that you should understand and avoid.

The most common way to format a string in Python is by using the % formatting operator. The predefined text template is provided on the left side of the operator in a *format string*. The values to insert into the template are provided as a single value or tuple of multiple values on the right side of the format operator. For example, here I use the % operator to convert difficult-to-read binary and hexadecimal values to integer strings:

```
a = 0b10111011
b = 0xc5f
print('Binary is %d, hex is %d' % (a, b))

>>>
Binary is 187, hex is 3167
```

The format string uses format specifiers (like %d) as placeholders that will be replaced by values from the right side of the formatting expression. The syntax for format specifiers comes from C's printf function, which has been inherited by Python (as well as by other programming languages). Python supports all the usual options you'd expect from printf, such as %s, %x, and %f format specifiers, as well as control over decimal places, padding, fill, and alignment. Many programmers who are new to Python start with C-style format strings because they're familiar and simple to use.

There are four problems with C-style format strings in Python.

The first problem is that if you change the type or order of data values in the tuple on the right side of a formatting expression, you can get errors due to type conversion incompatibility. For example, this simple formatting expression works:

```
key = 'my_var'
value = 1.234
formatted = '%-10s = %.2f' % (key, value)
print(formatted)

>>>
my_var     = 1.23
```

But if you swap key and value, you get an exception at runtime:

```
reordered_tuple = '%-10s = %.2f' % (value, key)

>>>
Traceback ...
TypeError: must be real number, not str
```

Similarly, leaving the right side parameters in the original order but changing the format string results in the same error:

```
reordered_string = '%.2f = %-10s' % (key, value)

>>>
Traceback ...
TypeError: must be real number, not str
```

To avoid this gotcha, you need to constantly check that the two sides of the % operator are in sync; this process is error prone because it must be done manually for every change.

The second problem with C-style formatting expressions is that they become difficult to read when you need to make small modifications to values before formatting them into a string—and this is an extremely common need. Here, I list the contents of my kitchen pantry without making inline changes:

```
pantry = [
    ('avocados', 1.25),
    ('bananas', 2.5),
    ('cherries', 15),
]
for i, (item, count) in enumerate(pantry):
    print('#%d: %-10s = %.2f' % (i, item, count))

>>>
#0: avocados   = 1.25
#1: bananas    = 2.50
#2: cherries   = 15.00
```

Now, I make a few modifications to the values that I'm formatting to make the printed message more useful. This causes the tuple in the formatting expression to become so long that it needs to be split across multiple lines, which hurts readability:

```
for i, (item, count) in enumerate(pantry):
    print('#%d: %-10s = %d' % (
        i + 1,
        item.title(),
        round(count)))
```

```
>>>
#1: Avocados   = 1
#2: Bananas    = 2
#3: Cherries   = 15
```

The third problem with formatting expressions is that if you want to use the same value in a format string multiple times, you have to repeat it in the right side tuple:

```
template = '%s loves food. See %s cook.'
name = 'Max'
formatted = template % (name, name)
print(formatted)
```

```
>>>
Max loves food. See Max cook.
```

This is especially annoying and error prone if you have to repeat small modifications to the values being formatted. For example, here I remembered to call the title() method multiple times, but I could have easily added the method call to one reference to name and not the other, which would cause mismatched output:

```
name = 'brad'
formatted = template % (name.title(), name.title())
print(formatted)
```

```
>>>
Brad loves food. See Brad cook.
```

To help solve some of these problems, the % operator in Python has the ability to also do formatting with a dictionary instead of a tuple. The keys from the dictionary are matched with format specifiers with the corresponding name, such as %(key)s. Here, I use this functionality to change the order of values on the right side of the formatting expression with no effect on the output, thus solving problem #1 from above:

```
key = 'my_var'
value = 1.234
```

```
old_way = '%-10s = %.2f' % (key, value)

new_way = '%(key)-10s = %(value).2f' % {
    'key': key, 'value': value}  # Original

reordered = '%(key)-10s = %(value).2f' % {
    'value': value, 'key': key}  # Swapped

assert old_way == new_way == reordered
```

Using dictionaries in formatting expressions also solves problem #3 from above by allowing multiple format specifiers to reference the same value, thus making it unnecessary to supply that value more than once:

```
name = 'Max'

template = '%s loves food. See %s cook.'
before = template % (name, name)    # Tuple

template = '%(name)s loves food. See %(name)s cook.'
after = template % {'name': name}  # Dictionary

assert before == after
```

However, dictionary format strings introduce and exacerbate other issues. For problem #2 above, regarding small modifications to values before formatting them, formatting expressions become longer and more visually noisy because of the presence of the dictionary key and colon operator on the right side. Here, I render the same string with and without dictionaries to show this problem:

```
for i, (item, count) in enumerate(pantry):
    before = '#%d: %-10s = %d' % (
        i + 1,
        item.title(),
        round(count))

    after = '#%(loop)d: %(item)-10s = %(count)d' % {
        'loop': i + 1,
        'item': item.title(),
        'count': round(count),
    }

    assert before == after
```

Using dictionaries in formatting expressions also increases verbosity, which is problem #4 with C-style formatting expressions in Python. Each key must be specified at least twice—once in the format specifier, once in the dictionary as a key, and potentially once more for the variable name that contains the dictionary value:

```
soup = 'lentil'
formatted = 'Today\'s soup is %(soup)s.' % {'soup': soup}
print(formatted)

>>>
Today's soup is lentil.
```

Besides the duplicative characters, this redundancy causes formatting expressions that use dictionaries to be long. These expressions often must span multiple lines, with the format strings being concatenated across multiple lines and the dictionary assignments having one line per value to use in formatting:

```
menu = {
    'soup': 'lentil',
    'oyster': 'kumamoto',
    'special': 'schnitzel',
}
template = ('Today\'s soup is %(soup)s, '
            'buy one get two %(oyster)s oysters, '
            'and our special entrée is %(special)s.')
formatted = template % menu
print(formatted)

>>>
Today's soup is lentil, buy one get two kumamoto oysters, and
➥our special entrée is schnitzel.
```

To understand what this formatting expression is going to produce, your eyes have to keep going back and forth between the lines of the format string and the lines of the dictionary. This disconnect makes it hard to spot bugs, and readability gets even worse if you need to make small modifications to any of the values before formatting.

There must be a better way.

The `format` Built-in and `str.format`

Python 3 added support for *advanced string formatting* that is more expressive than the old C-style format strings that use the % operator. For individual Python values, this new functionality can be accessed through the `format` built-in function. For example, here I use some of the new options (, for thousands separators and ∧ for centering) to format values:

```
a = 1234.5678
formatted = format(a, ',.2f')
print(formatted)

b = 'my string'
formatted = format(b, '^20s')
print('*', formatted, '*')

>>>
1,234.57
*        my string        *
```

You can use this functionality to format multiple values together by calling the new format method of the str type. Instead of using C-style format specifiers like %d, you can specify placeholders with {}. By default the placeholders in the format string are replaced by the corresponding positional arguments passed to the format method in the order in which they appear:

```
key = 'my_var'
value = 1.234

formatted = '{} = {}'.format(key, value)
print(formatted)

>>>
my_var = 1.234
```

Within each placeholder you can optionally provide a colon character followed by format specifiers to customize how values will be converted into strings (see help('FORMATTING') for the full range of options):

```
formatted = '{:<10} = {:.2f}'.format(key, value)
print(formatted)

>>>
my_var     = 1.23
```

The way to think about how this works is that the format specifiers will be passed to the format built-in function along with the value (format(value, '.2f') in the example above). The result of that function call is what replaces the placeholder in the overall formatted string. The formatting behavior can be customized per class using the __format__ special method.

With C-style format strings, you need to escape the % character (by doubling it) so it's not interpreted as a placeholder accidentally. With the `str.format` method you need to similarly escape braces:

```
print('%.2f%%' % 12.5)
print('{} replaces {{}}'.format(1.23))
```

```
>>>
12.50%
1.23 replaces {}
```

Within the braces you may also specify the positional index of an argument passed to the `format` method to use for replacing the placeholder. This allows the format string to be updated to reorder the output without requiring you to also change the right side of the formatting expression, thus addressing problem #1 from above:

```
formatted = '{1} = {0}'.format(key, value)
print(formatted)
```

```
>>>
1.234 = my_var
```

The same positional index may also be referenced multiple times in the format string without the need to pass the value to the `format` method more than once, which solves problem #3 from above:

```
formatted = '{0} loves food. See {0} cook.'.format(name)
print(formatted)
```

```
>>>
Max loves food. See Max cook.
```

Unfortunately, the new `format` method does nothing to address problem #2 from above, leaving your code difficult to read when you need to make small modifications to values before formatting them. There's little difference in readability between the old and new options, which are similarly noisy:

```
for i, (item, count) in enumerate(pantry):
    old_style = '#%d: %-10s = %d' % (
        i + 1,
        item.title(),
        round(count))

    new_style = '#{}: {:<10s} = {}'.format(
        i + 1,
        item.title(),
        round(count))

    assert old_style == new_style
```

There are even more advanced options for the specifiers used with the str.format method, such as using combinations of dictionary keys and list indexes in placeholders, and coercing values to Unicode and repr strings:

```
formatted = 'First letter is {menu[oyster][0]!r}'.format(
    menu=menu)
print(formatted)

>>>
First letter is 'k'
```

But these features don't help reduce the redundancy of repeated keys from problem #4 above. For example, here I compare the verbosity of using dictionaries in C-style formatting expressions to the new style of passing keyword arguments to the format method:

```
old_template = (
    'Today\'s soup is %(soup)s, '
    'buy one get two %(oyster)s oysters, '
    'and our special entrée is %(special)s.')
old_formatted = template % {
    'soup': 'lentil',
    'oyster': 'kumamoto',
    'special': 'schnitzel',
}

new_template = (
    'Today\'s soup is {soup}, '
    'buy one get two {oyster} oysters, '
    'and our special entrée is {special}.')
new_formatted = new_template.format(
    soup='lentil',
    oyster='kumamoto',
    special='schnitzel',
)

assert old_formatted == new_formatted
```

This style is slightly less noisy because it eliminates some quotes in the dictionary and a few characters in the format specifiers, but it's hardly compelling. Further, the advanced features of using dictionary keys and indexes within placeholders only provides a tiny subset of Python's expression functionality. This lack of expressiveness is so limiting that it undermines the value of the format method from str overall.

Given these shortcomings and the problems from C-style formatting expressions that remain (problems #2 and #4 from above), I suggest that you avoid the `str.format` method in general. It's important to know about the new mini language used in format specifiers (everything after the colon) and how to use the `format` built-in function. But the rest of the `str.format` method should be treated as a historical artifact to help you understand how Python's new *f-strings* work and why they're so great.

Interpolated Format Strings

Python 3.6 added *interpolated format strings—f-strings* for short—to solve these issues once and for all. This new language syntax requires you to prefix format strings with an f character, which is similar to how byte strings are prefixed with a b character and raw (unescaped) strings are prefixed with an r character.

F-strings take the expressiveness of format strings to the extreme, solving problem #4 from above by completely eliminating the redundancy of providing keys and values to be formatted. They achieve this pithiness by allowing you to reference all names in the current Python scope as part of a formatting expression:

```
key = 'my_var'
value = 1.234

formatted = f'{key} = {value}'
print(formatted)

>>>
my_var = 1.234
```

All of the same options from the new `format` built-in mini language are available after the colon in the placeholders within an f-string, as is the ability to coerce values to Unicode and `repr` strings similar to the `str.format` method:

```
formatted = f'{key!r:<10} = {value:.2f}'
print(formatted)

>>>
'my_var'   = 1.23
```

Formatting with f-strings is shorter than using C-style format strings with the % operator and the `str.format` method in all cases. Here, I show all these options together in order of shortest to longest, and line up the left side of the assignment so you can easily compare them:

```
f_string = f'{key:<10} = {value:.2f}'

c_tuple  = '%-10s = %.2f' % (key, value)

str_args = '{:<10} = {:.2f}'.format(key, value)

str_kw   = '{key:<10} = {value:.2f}'.format(key=key,
                                            value=value)

c_dict   = '%(key)-10s = %(value).2f' % {'key': key,
                                         'value': value}

assert c_tuple == c_dict == f_string
assert str_args == str_kw == f_string
```

F-strings also enable you to put a full Python expression within the placeholder braces, solving problem #2 from above by allowing small modifications to the values being formatted with concise syntax. What took multiple lines with C-style formatting and the str.format method now easily fits on a single line:

```
for i, (item, count) in enumerate(pantry):
    old_style = '#%d: %-10s = %d' % (
        i + 1,
        item.title(),
        round(count))

    new_style = '#{}: {:<10s} = {}'.format(
        i + 1,
        item.title(),
        round(count))

    f_string = f'#{i+1}: {item.title():<10s} = {round(count)}'

    assert old_style == new_style == f_string
```

Or, if it's clearer, you can split an f-string over multiple lines by relying on adjacent-string concatenation (similar to C). Even though this is longer than the single-line version, it's still much clearer than any of the other multiline approaches:

```
for i, (item, count) in enumerate(pantry):
    print(f'#{i+1}: '
```

```
    f'{item.title():<10s} = '
    f'{round(count)}')
```

```
>>>
#1: Avocados   = 1
#2: Bananas    = 2
#3: Cherries   = 15
```

Python expressions may also appear within the format specifier options. For example, here I parameterize the number of digits to print by using a variable instead of hard-coding it in the format string:

```
places = 3
number = 1.23456
print(f'My number is {number:.{places}f}')
```

```
>>>
My number is 1.235
```

The combination of expressiveness, terseness, and clarity provided by f-strings makes them the best built-in option for Python programmers. Any time you find yourself needing to format values into strings, choose f-strings over the alternatives.

Things to Remember

✦ C-style format strings that use the % operator suffer from a variety of gotchas and verbosity problems.

✦ The str.format method introduces some useful concepts in its formatting specifiers mini language, but it otherwise repeats the mistakes of C-style format strings and should be avoided.

✦ F-strings are a new syntax for formatting values into strings that solves the biggest problems with C-style format strings.

✦ F-strings are succinct yet powerful because they allow for arbitrary Python expressions to be directly embedded within format specifiers.

Item 5: Write Helper Functions Instead of Complex Expressions

Python's pithy syntax makes it easy to write single-line expressions that implement a lot of logic. For example, say that I want to decode the query string from a URL. Here, each query string parameter represents an integer value:

```
from urllib.parse import parse_qs
```

```
my_values = parse_qs('red=5&blue=0&green=',
                     keep_blank_values=True)
print(repr(my_values))

>>>
{'red': ['5'], 'blue': ['0'], 'green': ['']}
```

Some query string parameters may have multiple values, some may have single values, some may be present but have blank values, and some may be missing entirely. Using the get method on the result dictionary will return different values in each circumstance:

```
print('Red:     ', my_values.get('red'))
print('Green:   ', my_values.get('green'))
print('Opacity: ', my_values.get('opacity'))

>>>
Red:      ['5']
Green:    ['']
Opacity:  None
```

It'd be nice if a default value of 0 were assigned when a parameter isn't supplied or is blank. I might choose to do this with Boolean expressions because it feels like this logic doesn't merit a whole if statement or helper function quite yet.

Python's syntax makes this choice all too easy. The trick here is that the empty string, the empty list, and zero all evaluate to False implicitly. Thus, the expressions below will evaluate to the subexpression after the or operator when the first subexpression is False:

```
# For query string 'red=5&blue=0&green='
red = my_values.get('red', [''])[0] or 0
green = my_values.get('green', [''])[0] or 0
opacity = my_values.get('opacity', [''])[0] or 0
print(f'Red:     {red!r}')
print(f'Green:   {green!r}')
print(f'Opacity: {opacity!r}')

>>>
Red:      '5'
Green:    0
Opacity:  0
```

The red case works because the key is present in the my_values dictionary. The value is a list with one member: the string '5'. This string implicitly evaluates to True, so red is assigned to the first part of the or expression.

The green case works because the value in the my_values dictionary is a list with one member: an empty string. The empty string implicitly evaluates to False, causing the or expression to evaluate to 0.

The opacity case works because the value in the my_values dictionary is missing altogether. The behavior of the get method is to return its second argument if the key doesn't exist in the dictionary (see Item 16: "Prefer get Over in and KeyError to Handle Missing Dictionary Keys"). The default value in this case is a list with one member: an empty string. When opacity isn't found in the dictionary, this code does exactly the same thing as the green case.

However, this expression is difficult to read, and it still doesn't do everything I need. I'd also want to ensure that all the parameter values are converted to integers so I can immediately use them in mathematical expressions. To do that, I'd wrap each expression with the int built-in function to parse the string as an integer:

```
red = int(my_values.get('red', [''])[0] or 0)
```

This is now extremely hard to read. There's so much visual noise. The code isn't approachable. A new reader of the code would have to spend too much time picking apart the expression to figure out what it actually does. Even though it's nice to keep things short, it's not worth trying to fit this all on one line.

Python has if/else conditional—or ternary—expressions to make cases like this clearer while keeping the code short:

```
red_str = my_values.get('red', [''])
red = int(red_str[0]) if red_str[0] else 0
```

This is better. For less complicated situations, if/else conditional expressions can make things very clear. But the example above is still not as clear as the alternative of a full if/else statement over multiple lines. Seeing all of the logic spread out like this makes the dense version seem even more complex:

```
green_str = my_values.get('green', [''])
if green_str[0]:
    green = int(green_str[0])
else:
    green = 0
```

If you need to reuse this logic repeatedly—even just two or three times, as in this example—then writing a helper function is the way to go:

```
def get_first_int(values, key, default=0):
    found = values.get(key, [''])
```

```
    if found[0]:
        return int(found[0])
    return default
```

The calling code is much clearer than the complex expression using `or` and the two-line version using the `if/else` expression:

```
green = get_first_int(my_values, 'green')
```

As soon as expressions get complicated, it's time to consider splitting them into smaller pieces and moving logic into helper functions. What you gain in readability always outweighs what brevity may have afforded you. Avoid letting Python's pithy syntax for complex expressions from getting you into a mess like this. Follow the *DRY principle*: Don't repeat yourself.

Things to Remember

✦ Python's syntax makes it easy to write single-line expressions that are overly complicated and difficult to read.

✦ Move complex expressions into helper functions, especially if you need to use the same logic repeatedly.

✦ An `if/else` expression provides a more readable alternative to using the Boolean operators `or` and `and` in expressions.

Item 6: Prefer Multiple Assignment Unpacking Over Indexing

Python has a built-in `tuple` type that can be used to create immutable, ordered sequences of values. In the simplest case, a `tuple` is a pair of two values, such as keys and values from a dictionary:

```
snack_calories = {
    'chips': 140,
    'popcorn': 80,
    'nuts': 190,
}
items = tuple(snack_calories.items())
print(items)

>>>
(('chips', 140), ('popcorn', 80), ('nuts', 190))
```

The values in tuples can be accessed through numerical indexes:

```
item = ('Peanut butter', 'Jelly')
first = item[0]
second = item[1]
print(first, 'and', second)
```

```
>>>
Peanut butter and Jelly
```

Once a tuple is created, you can't modify it by assigning a new value to an index:

```
pair = ('Chocolate', 'Peanut butter')
pair[0] = 'Honey'
```

```
>>>
Traceback ...
TypeError: 'tuple' object does not support item assignment
```

Python also has syntax for *unpacking*, which allows for assigning multiple values in a single statement. The patterns that you specify in unpacking assignments look a lot like trying to mutate tuples—which isn't allowed—but they actually work quite differently. For example, if you know that a tuple is a pair, instead of using indexes to access its values, you can assign it to a tuple of two variable names:

```
item = ('Peanut butter', 'Jelly')
first, second = item  # Unpacking
print(first, 'and', second)
```

```
>>>
Peanut butter and Jelly
```

Unpacking has less visual noise than accessing the tuple's indexes, and it often requires fewer lines. The same pattern matching syntax of unpacking works when assigning to lists, sequences, and multiple levels of arbitrary iterables within iterables. I don't recommend doing the following in your code, but it's important to know that it's possible and how it works:

```
favorite_snacks = {
    'salty': ('pretzels', 100),
    'sweet': ('cookies', 180),
    'veggie': ('carrots', 20),
}

((type1, (name1, cals1)),
 (type2, (name2, cals2)),
 (type3, (name3, cals3))) = favorite_snacks.items()
```

```
print(f'Favorite {type1} is {name1} with {cals1} calories')
print(f'Favorite {type2} is {name2} with {cals2} calories')
print(f'Favorite {type3} is {name3} with {cals3} calories')

>>>
Favorite salty is pretzels with 100 calories
Favorite sweet is cookies with 180 calories
Favorite veggie is carrots with 20 calories
```

Newcomers to Python may be surprised to learn that unpacking can even be used to swap values in place without the need to create temporary variables. Here, I use typical syntax with indexes to swap the values between two positions in a list as part of an ascending order sorting algorithm:

```
def bubble_sort(a):
    for _ in range(len(a)):
        for i in range(1, len(a)):
            if a[i] < a[i-1]:
                temp = a[i]
                a[i] = a[i-1]
                a[i-1] = temp

names = ['pretzels', 'carrots', 'arugula', 'bacon']
bubble_sort(names)
print(names)

>>>
['arugula', 'bacon', 'carrots', 'pretzels']
```

However, with unpacking syntax, it's possible to swap indexes in a single line:

```
def bubble_sort(a):
    for _ in range(len(a)):
        for i in range(1, len(a)):
            if a[i] < a[i-1]:
                a[i-1], a[i] = a[i], a[i-1]  # Swap

names = ['pretzels', 'carrots', 'arugula', 'bacon']
bubble_sort(names)
print(names)

>>>
['arugula', 'bacon', 'carrots', 'pretzels']
```

The way this swap works is that the right side of the assignment (a[i], a[i-1]) is evaluated first, and its values are put into a new temporary, unnamed tuple (such as ('carrots', 'pretzels') on the first iteration of the loops). Then, the unpacking pattern from the left side of the assignment (a[i-1], a[i]) is used to receive that tuple value and assign it to the variable names a[i-1] and a[i], respectively. This replaces 'pretzels' with 'carrots' at index 0 and 'carrots' with 'pretzels' at index 1. Finally, the temporary unnamed tuple silently goes away.

Another valuable application of unpacking is in the target list of for loops and similar constructs, such as comprehensions and generator expressions (see Item 27: "Use Comprehensions Instead of map and filter" for those). As an example for contrast, here I iterate over a list of snacks without using unpacking:

```
snacks = [('bacon', 350), ('donut', 240), ('muffin', 190)]
for i in range(len(snacks)):
    item = snacks[i]
    name = item[0]
    calories = item[1]
    print(f'#{i+1}: {name} has {calories} calories')
```

```
>>>
#1: bacon has 350 calories
#2: donut has 240 calories
#3: muffin has 190 calories
```

This works, but it's noisy. There are a lot of extra characters required in order to index into the various levels of the snacks structure. Here, I achieve the same output by using unpacking along with the enumerate built-in function (see Item 7: "Prefer enumerate Over range"):

```
for rank, (name, calories) in enumerate(snacks, 1):
    print(f'#{rank}: {name} has {calories} calories')
```

```
>>>
#1: bacon has 350 calories
#2: donut has 240 calories
#3: muffin has 190 calories
```

This is the Pythonic way to write this type of loop; it's short and easy to understand. There's usually no need to access anything using indexes.

Python provides additional unpacking functionality for list construction (see Item 13: "Prefer Catch-All Unpacking Over Slicing"), function arguments (see Item 22: "Reduce Visual Noise with Variable Positional Arguments"), keyword arguments (see Item 23: "Provide Optional Behavior with Keyword Arguments"), multiple return values (see Item 19: "Never Unpack More Than Three Variables When Functions Return Multiple Values"), and more.

Using unpacking wisely will enable you to avoid indexing when possible, resulting in clearer and more Pythonic code.

Things to Remember

✦ Python has special syntax called unpacking for assigning multiple values in a single statement.

✦ Unpacking is generalized in Python and can be applied to any iterable, including many levels of iterables within iterables.

✦ Reduce visual noise and increase code clarity by using unpacking to avoid explicitly indexing into sequences.

Item 7: Prefer enumerate Over range

The range built-in function is useful for loops that iterate over a set of integers:

```
from random import randint

random_bits = 0
for i in range(32):
    if randint(0, 1):
        random_bits |= 1 << i


print(bin(random_bits))
```

```
>>>
0b11101000100100000111000010000001
```

When you have a data structure to iterate over, like a list of strings, you can loop directly over the sequence:

```
flavor_list = ['vanilla', 'chocolate', 'pecan', 'strawberry']
for flavor in flavor_list:
    print(f'{flavor} is delicious')
```

```
>>>
vanilla is delicious
chocolate is delicious
pecan is delicious
strawberry is delicious
```

Often, you'll want to iterate over a list and also know the index of the current item in the list. For example, say that I want to print the ranking of my favorite ice cream flavors. One way to do it is by using range:

```
for i in range(len(flavor_list)):
    flavor = flavor_list[i]
    print(f'{i + 1}: {flavor}')
```

```
>>>
1: vanilla
2: chocolate
3: pecan
4: strawberry
```

This looks clumsy compared with the other examples of iterating over flavor_list or range. I have to get the length of the list. I have to index into the array. The multiple steps make it harder to read.

Python provides the enumerate built-in function to address this situation. enumerate wraps any iterator with a lazy generator (see Item 30: "Consider Generators Instead of Returning Lists"). enumerate yields pairs of the loop index and the next value from the given iterator. Here, I manually advance the returned iterator with the next built-in function to demonstrate what it does:

```
it = enumerate(flavor_list)
print(next(it))
print(next(it))
```

```
>>>
(0, 'vanilla')
(1, 'chocolate')
```

Each pair yielded by enumerate can be succinctly unpacked in a for statement (see Item 6: "Prefer Multiple Assignment Unpacking Over Indexing" for how that works). The resulting code is much clearer:

```
for i, flavor in enumerate(flavor_list):
    print(f'{i + 1}: {flavor}')
```

```
>>>
1: vanilla
2: chocolate
3: pecan
4: strawberry
```

I can make this even shorter by specifying the number from which enumerate should begin counting (1 in this case) as the second parameter:

```
for i, flavor in enumerate(flavor_list, 1):
    print(f'{i}: {flavor}')
```

Things to Remember

✦ enumerate provides concise syntax for looping over an iterator and getting the index of each item from the iterator as you go.

✦ Prefer enumerate instead of looping over a range and indexing into a sequence.

✦ You can supply a second parameter to enumerate to specify the number from which to begin counting (zero is the default).

Item 8: Use `zip` to Process Iterators in Parallel

Often in Python you find yourself with many lists of related objects. List comprehensions make it easy to take a source `list` and get a derived `list` by applying an expression (see Item 27: "Use Comprehensions Instead of `map` and `filter`"):

```
names = ['Cecilia', 'Lise', 'Marie']
counts = [len(n) for n in names]
print(counts)
```

```
>>>
[7, 4, 5]
```

The items in the derived `list` are related to the items in the source `list` by their indexes. To iterate over both lists in parallel, I can iterate over the length of the `names` source `list`:

```
longest_name = None
max_count = 0

for i in range(len(names)):
    count = counts[i]
    if count > max_count:
        longest_name = names[i]
        max_count = count


print(longest_name)
```

```
>>>
Cecilia
```

The problem is that this whole loop statement is visually noisy. The indexes into `names` and `counts` make the code hard to read. Indexing into the arrays by the loop index i happens twice. Using enumerate (see Item 7: "Prefer enumerate Over range") improves this slightly, but it's still not ideal:

```
for i, name in enumerate(names):
    count = counts[i]
    if count > max_count:
        longest_name = name
        max_count = count
```

To make this code clearer, Python provides the zip built-in function. zip wraps two or more iterators with a lazy generator. The zip generator yields tuples containing the next value from each iterator. These tuples can be unpacked directly within a for statement (see Item 6: "Prefer Multiple Assignment Unpacking Over Indexing"). The resulting code is much cleaner than the code for indexing into multiple lists:

```
for name, count in zip(names, counts):
    if count > max_count:
        longest_name = name
        max_count = count
```

zip consumes the iterators it wraps one item at a time, which means it can be used with infinitely long inputs without risk of a program using too much memory and crashing.

However, beware of zip's behavior when the input iterators are of different lengths. For example, say that I add another item to names above but forget to update counts. Running zip on the two input lists will have an unexpected result:

```
names.append('Rosalind')
for name, count in zip(names, counts):
    print(name)

>>>
Cecilia
Lise
Marie
```

The new item for 'Rosalind' isn't there. Why not? This is just how zip works. It keeps yielding tuples until any one of the wrapped iterators is exhausted. Its output is as long as its shortest input. This approach works fine when you know that the iterators are of the same length, which is often the case for derived lists created by list comprehensions.

But in many other cases, the truncating behavior of zip is surprising and bad. If you don't expect the lengths of the lists passed to zip to be equal, consider using the zip_longest function from the itertools built-in module instead:

```
import itertools

for name, count in itertools.zip_longest(names, counts):
    print(f'{name}: {count}')
```

```
>>>
Cecilia: 7
Lise: 4
Marie: 5
Rosalind: None
```

zip_longest replaces missing values—the length of the string 'Rosalind' in this case—with whatever fillvalue is passed to it, which defaults to None.

Things to Remember

✦ The zip built-in function can be used to iterate over multiple iterators in parallel.

✦ zip creates a lazy generator that produces tuples, so it can be used on infinitely long inputs.

✦ zip truncates its output silently to the shortest iterator if you supply it with iterators of different lengths.

✦ Use the zip_longest function from the itertools built-in module if you want to use zip on iterators of unequal lengths without truncation.

Item 9: Avoid else Blocks After for and while Loops

Python loops have an extra feature that is not available in most other programming languages: You can put an else block immediately after a loop's repeated interior block:

```
for i in range(3):
    print('Loop', i)
else:
    print('Else block!')
```

```
>>>
Loop 0
Loop 1
Loop 2
Else block!
```

Surprisingly, the else block runs immediately after the loop finishes. Why is the clause called "else"? Why not "and"? In an if/else statement, else means "Do this if the block before this doesn't happen." In a try/except statement, except has the same definition: "Do this if trying the block before this failed."

Similarly, else from try/except/else follows this pattern (see Item 65: "Take Advantage of Each Block in try/except/else/finally") because it means "Do this if there was no exception to handle." try/finally is also intuitive because it means "Always do this after trying the block before."

Given all the uses of else, except, and finally in Python, a new programmer might assume that the else part of for/else means "Do this if the loop wasn't completed." In reality, it does exactly the opposite. Using a break statement in a loop actually skips the else block:

```
for i in range(3):
    print('Loop', i)
    if i == 1:
        break
else:
    print('Else block!')

>>>
Loop 0
Loop 1
```

Another surprise is that the else block runs immediately if you loop over an empty sequence:

```
for x in []:
    print('Never runs')
else:
    print('For Else block!')

>>>
For Else block!
```

The else block also runs when while loops are initially False:

```
while False:
    print('Never runs')
else:
    print('While Else block!')

>>>
While Else block!
```

The rationale for these behaviors is that else blocks after loops are useful when using loops to search for something. For example, say that I want to determine whether two numbers are coprime (that is, their only common divisor is 1). Here, I iterate through every possible common divisor and test the numbers. After every option has been tried, the loop ends. The `else` block runs when the numbers are coprime because the loop doesn't encounter a break:

```
a = 4
b = 9

for i in range(2, min(a, b) + 1):
    print('Testing', i)
    if a % i == 0 and b % i == 0:
        print('Not coprime')
        break
else:
    print('Coprime')

>>>
Testing 2
Testing 3
Testing 4
Coprime
```

In practice, I wouldn't write the code this way. Instead, I'd write a helper function to do the calculation. Such a helper function is written in two common styles.

The first approach is to return early when I find the condition I'm looking for. I return the default outcome if I fall through the loop:

```
def coprime(a, b):
    for i in range(2, min(a, b) + 1):
        if a % i == 0 and b % i == 0:
            return False
    return True

assert coprime(4, 9)
assert not coprime(3, 6)
```

The second way is to have a result variable that indicates whether I've found what I'm looking for in the loop. I break out of the loop as soon as I find something:

```
def coprime_alternate(a, b):
    is_coprime = True
    for i in range(2, min(a, b) + 1):
        if a % i == 0 and b % i == 0:
            is_coprime = False
            break
    return is_coprime
```

```
assert coprime_alternate(4, 9)
assert not coprime_alternate(3, 6)
```

Both approaches are much clearer to readers of unfamiliar code. Depending on the situation, either may be a good choice. However, the expressivity you gain from the else block doesn't outweigh the burden you put on people (including yourself) who want to understand your code in the future. Simple constructs like loops should be self-evident in Python. You should avoid using else blocks after loops entirely.

Things to Remember

- ✦ Python has special syntax that allows else blocks to immediately follow for and while loop interior blocks.

- ✦ The else block after a loop runs only if the loop body did not encounter a break statement.

- ✦ Avoid using else blocks after loops because their behavior isn't intuitive and can be confusing.

Item 10: Prevent Repetition with Assignment Expressions

An assignment expression—also known as the *walrus operator*—is a new syntax introduced in Python 3.8 to solve a long-standing problem with the language that can cause code duplication. Whereas normal assignment statements are written a = b and pronounced "a equals b," these assignments are written a := b and pronounced "a *walrus* b" (because := looks like a pair of eyeballs and tusks).

Assignment expressions are useful because they enable you to assign variables in places where assignment statements are disallowed, such as in the conditional expression of an if statement. An assignment expression's value evaluates to whatever was assigned to the identifier on the left side of the walrus operator.

For example, say that I have a basket of fresh fruit that I'm trying to manage for a juice bar. Here, I define the contents of the basket:

```
fresh_fruit = {
    'apple': 10,
    'banana': 8,
    'lemon': 5,
}
```

When a customer comes to the counter to order some lemonade, I need to make sure there is at least one lemon in the basket to squeeze. Here, I do this by retrieving the count of lemons and then using an if statement to check for a non-zero value:

```
def make_lemonade(count):
    ...

def out_of_stock():
    ...

count = fresh_fruit.get('lemon', 0)
if count:
    make_lemonade(count)
else:
    out_of_stock()
```

The problem with this seemingly simple code is that it's noisier than it needs to be. The count variable is used only within the first block of the if statement. Defining count above the if statement causes it to appear to be more important than it really is, as if all code that follows, including the else block, will need to access the count variable, when in fact that is not the case.

This pattern of fetching a value, checking to see if it's non-zero, and then using it is extremely common in Python. Many programmers try to work around the multiple references to count with a variety of tricks that hurt readability (see Item 5: "Write Helper Functions Instead of Complex Expressions" for an example). Luckily, assignment expressions were added to the language to streamline exactly this type of code. Here, I rewrite this example using the walrus operator:

```
if count := fresh_fruit.get('lemon', 0):
    make_lemonade(count)
else:
    out_of_stock()
```

Though this is only one line shorter, it's a lot more readable because it's now clear that count is only relevant to the first block of the if statement. The assignment expression is first assigning a value to the count variable, and then evaluating that value in the context of the if statement to determine how to proceed with flow control. This two-step behavior—assign and then evaluate—is the fundamental nature of the walrus operator.

Lemons are quite potent, so only one is needed for my lemonade recipe, which means a non-zero check is good enough. If a customer orders a cider, though, I need to make sure that I have at least four apples. Here, I do this by fetching the count from the fruit_basket dictionary, and then using a comparison in the if statement conditional expression:

```
def make_cider(count):
    ...

count = fresh_fruit.get('apple', 0)
if count >= 4:
    make_cider(count)
else:
    out_of_stock()
```

This has the same problem as the lemonade example, where the assignment of count puts distracting emphasis on that variable. Here, I improve the clarity of this code by also using the walrus operator:

```
if (count := fresh_fruit.get('apple', 0)) >= 4:
    make_cider(count)
else:
    out_of_stock()
```

This works as expected and makes the code one line shorter. It's important to note how I needed to surround the assignment expression with parentheses to compare it with 4 in the if statement. In the lemonade example, no surrounding parentheses were required because the assignment expression stood on its own as a non-zero check; it wasn't a subexpression of a larger expression. As with other expressions, you should avoid surrounding assignment expressions with parentheses when possible.

Another common variation of this repetitive pattern occurs when I need to assign a variable in the enclosing scope depending on some condition, and then reference that variable shortly afterward in a function call. For example, say that a customer orders some banana smoothies. In order to make them, I need to have at least two bananas' worth of slices, or else an OutOfBananas exception will be raised. Here, I implement this logic in a typical way:

```
def slice_bananas(count):
    ...

class OutOfBananas(Exception):
    pass
```

```
def make_smoothies(count):
    ...

pieces = 0
count = fresh_fruit.get('banana', 0)
if count >= 2:
    pieces = slice_bananas(count)

try:
    smoothies = make_smoothies(pieces)
except OutOfBananas:
    out_of_stock()
```

The other common way to do this is to put the pieces = 0 assignment in the else block:

```
count = fresh_fruit.get('banana', 0)
if count >= 2:
    pieces = slice_bananas(count)
else:
    pieces = 0

try:
    smoothies = make_smoothies(pieces)
except OutOfBananas:
    out_of_stock()
```

This second approach can feel odd because it means that the pieces variable has two different locations—in each block of the if statement—where it can be initially defined. This split definition technically works because of Python's scoping rules (see Item 21: "Know How Closures Interact with Variable Scope"), but it isn't easy to read or discover, which is why many people prefer the construct above, where the pieces = 0 assignment is first.

The walrus operator can again be used to shorten this example by one line of code. This small change removes any emphasis on the count variable. Now, it's clearer that pieces will be important beyond the if statement:

```
pieces = 0
if (count := fresh_fruit.get('banana', 0)) >= 2:
    pieces = slice_bananas(count)

try:
    smoothies = make_smoothies(pieces)
except OutOfBananas:
    out_of_stock()
```

Using the walrus operator also improves the readability of splitting the definition of pieces across both parts of the if statement. It's easier to trace the pieces variable when the count definition no longer precedes the if statement:

```
if (count := fresh_fruit.get('banana', 0)) >= 2:
    pieces = slice_bananas(count)
else:
    pieces = 0

try:
    smoothies = make_smoothies(pieces)
except OutOfBananas:
    out_of_stock()
```

One frustration that programmers who are new to Python often have is the lack of a flexible switch/case statement. The general style for approximating this type of functionality is to have a deep nesting of multiple if, elif, and else statements.

For example, imagine that I want to implement a system of precedence so that each customer automatically gets the best juice available and doesn't have to order. Here, I define logic to make it so banana smoothies are served first, followed by apple cider, and then finally lemonade:

```
count = fresh_fruit.get('banana', 0)
if count >= 2:
    pieces = slice_bananas(count)
    to_enjoy = make_smoothies(pieces)
else:
    count = fresh_fruit.get('apple', 0)
    if count >= 4:
        to_enjoy = make_cider(count)
    else:
        count = fresh_fruit.get('lemon', 0)
        if count:
            to_enjoy = make_lemonade(count)
        else:
            to_enjoy = 'Nothing'
```

Ugly constructs like this are surprisingly common in Python code. Luckily, the walrus operator provides an elegant solution that can feel nearly as versatile as dedicated syntax for switch/case statements:

```
if (count := fresh_fruit.get('banana', 0)) >= 2:
    pieces = slice_bananas(count)
    to_enjoy = make_smoothies(pieces)
```

```
elif (count := fresh_fruit.get('apple', 0)) >= 4:
    to_enjoy = make_cider(count)
elif count := fresh_fruit.get('lemon', 0):
    to_enjoy = make_lemonade(count)
else:
    to_enjoy = 'Nothing'
```

The version that uses assignment expressions is only five lines shorter than the original, but the improvement in readability is vast due to the reduction in nesting and indentation. If you ever see such ugly constructs emerge in your code, I suggest that you move them over to using the walrus operator if possible.

Another common frustration of new Python programmers is the lack of a do/while loop construct. For example, say that I want to bottle juice as new fruit is delivered until there's no fruit remaining. Here, I implement this logic with a while loop:

```
def pick_fruit():
    ...

def make_juice(fruit, count):
    ...

bottles = []
fresh_fruit = pick_fruit()
while fresh_fruit:
    for fruit, count in fresh_fruit.items():
        batch = make_juice(fruit, count)
        bottles.extend(batch)
    fresh_fruit = pick_fruit()
```

This is repetitive because it requires two separate fresh_fruit = pick_fruit() calls: one before the loop to set initial conditions, and another at the end of the loop to replenish the list of delivered fruit.

A strategy for improving code reuse in this situation is to use the *loop-and-a-half* idiom. This eliminates the redundant lines, but it also undermines the while loop's contribution by making it a dumb infinite loop. Now, all of the flow control of the loop depends on the conditional break statement:

```
bottles = []
while True:                        # Loop
    fresh_fruit = pick_fruit()
    if not fresh_fruit:            # And a half
        break
```

```
for fruit, count in fresh_fruit.items():
    batch = make_juice(fruit, count)
    bottles.extend(batch)
```

The walrus operator obviates the need for the loop-and-a-half idiom by allowing the `fresh_fruit` variable to be reassigned and then conditionally evaluated each time through the `while` loop. This solution is short and easy to read, and it should be the preferred approach in your code:

```
bottles = []
while fresh_fruit := pick_fruit():
    for fruit, count in fresh_fruit.items():
        batch = make_juice(fruit, count)
        bottles.extend(batch)
```

There are many other situations where assignment expressions can be used to eliminate redundancy (see Item 29: "Avoid Repeated Work in Comprehensions by Using Assignment Expressions" for another). In general, when you find yourself repeating the same expression or assignment multiple times within a grouping of lines, it's time to consider using assignment expressions in order to improve readability.

Things to Remember

✦ Assignment expressions use the walrus operator (:=) to both assign and evaluate variable names in a single expression, thus reducing repetition.

✦ When an assignment expression is a subexpression of a larger expression, it must be surrounded with parentheses.

✦ Although switch/case statements and do/while loops are not available in Python, their functionality can be emulated much more clearly by using assignment expressions.

Chapter 2

Lists and Dictionaries

Many programs are written to automate repetitive tasks that are better suited to machines than to humans. In Python, the most common way to organize this kind of work is by using a sequence of values stored in a list type. Lists are extremely versatile and can be used to solve a variety of problems.

A natural complement to lists is the dict type, which stores lookup keys mapped to corresponding values (in what is often called an *associative array* or a *hash table*). Dictionaries provide constant time (amortized) performance for assignments and accesses, which means they are ideal for bookkeeping dynamic information.

Python has special syntax and built-in modules that enhance readability and extend the capabilities of lists and dictionaries beyond what you might expect from simple array, vector, and hash table types in other languages.

Item 11: Know How to Slice Sequences

Python includes syntax for *slicing* sequences into pieces. Slicing allows you to access a subset of a sequence's items with minimal effort. The simplest uses for slicing are the built-in types list, str, and bytes. Slicing can be extended to any Python class that implements the __getitem__ and __setitem__ special methods (see Item 43: "Inherit from collections.abc for Custom Container Types").

The basic form of the slicing syntax is somelist[start:end], where start is inclusive and end is exclusive:

```
a = ['a', 'b', 'c', 'd', 'e', 'f', 'g', 'h']
print('Middle two:   ', a[3:5])
print('All but ends:', a[1:7])

>>>
Middle two:    ['d', 'e']
All but ends: ['b', 'c', 'd', 'e', 'f', 'g']
```

When slicing from the start of a list, you should leave out the zero index to reduce visual noise:

```
assert a[:5] == a[0:5]
```

When slicing to the end of a list, you should leave out the final index because it's redundant:

```
assert a[5:] == a[5:len(a)]
```

Using negative numbers for slicing is helpful for doing offsets relative to the end of a list. All of these forms of slicing would be clear to a new reader of your code:

```
a[:]       # ['a', 'b', 'c', 'd', 'e', 'f', 'g', 'h']
a[:5]      # ['a', 'b', 'c', 'd', 'e']
a[:-1]     # ['a', 'b', 'c', 'd', 'e', 'f', 'g']
a[4:]      #                         ['e', 'f', 'g', 'h']
a[-3:]     #                              ['f', 'g', 'h']
a[2:5]     #           ['c', 'd', 'e']
a[2:-1]    #           ['c', 'd', 'e', 'f', 'g']
a[-3:-1]   #                              ['f', 'g']
```

There are no surprises here, and I encourage you to use these variations.

Slicing deals properly with start and end indexes that are beyond the boundaries of a list by silently omitting missing items. This behavior makes it easy for your code to establish a maximum length to consider for an input sequence:

```
first_twenty_items = a[:20]
last_twenty_items = a[-20:]
```

In contrast, accessing the same index directly causes an exception:

```
a[20]
```

```
>>>
Traceback ...
IndexError: list index out of range
```

> **Note**
>
> Beware that indexing a list by a negated variable is one of the few situations in which you can get surprising results from slicing. For example, the expression somelist[-n:] will work fine when n is greater than one (e.g., somelist[-3:]). However, when n is zero, the expression somelist[-0:] is equivalent to somelist[:] and will result in a copy of the original list.

The result of slicing a list is a whole new list. References to the objects from the original list are maintained. Modifying the result of slicing won't affect the original list:

```
b = a[3:]
print('Before:    ', b)
b[1] = 99
print('After:     ', b)
print('No change:', a)

>>>
Before:     ['d', 'e', 'f', 'g', 'h']
After:      ['d', 99, 'f', 'g', 'h']
No change: ['a', 'b', 'c', 'd', 'e', 'f', 'g', 'h']
```

When used in assignments, slices replace the specified range in the original list. Unlike unpacking assignments (such as a, b = c[:2]; see Item 6: "Prefer Multiple Assignment Unpacking Over Indexing"), the lengths of slice assignments don't need to be the same. The values before and after the assigned slice will be preserved. Here, the list shrinks because the replacement list is shorter than the specified slice:

```
print('Before ', a)
a[2:7] = [99, 22, 14]
print('After  ', a)

>>>
Before  ['a', 'b', 'c', 'd', 'e', 'f', 'g', 'h']
After   ['a', 'b', 99, 22, 14, 'h']
```

And here the list grows because the assigned list is longer than the specific slice:

```
print('Before ', a)
a[2:3] = [47, 11]
print('After  ', a)

>>>
Before  ['a', 'b', 99, 22, 14, 'h']
After   ['a', 'b', 47, 11, 22, 14, 'h']
```

If you leave out both the start and the end indexes when slicing, you end up with a copy of the original list:

```
b = a[:]
assert b == a and b is not a
```

If you assign to a slice with no start or end indexes, you replace the entire contents of the list with a copy of what's referenced (instead of allocating a new list):

```
b = a
print('Before a', a)
print('Before b', b)
a[:] = [101, 102, 103]
assert a is b              # Still the same list object
print('After a ', a)       # Now has different contents
print('After b ', b)       # Same list, so same contents as a

>>>
Before a ['a', 'b', 47, 11, 22, 14, 'h']
Before b ['a', 'b', 47, 11, 22, 14, 'h']
After a  [101, 102, 103]
After b  [101, 102, 103]
```

Things to Remember

✦ Avoid being verbose when slicing: Don't supply 0 for the start index or the length of the sequence for the end index.

✦ Slicing is forgiving of start or end indexes that are out of bounds, which means it's easy to express slices on the front or back boundaries of a sequence (like a[:20] or a[-20:]).

✦ Assigning to a list slice replaces that range in the original sequence with what's referenced even if the lengths are different.

Item 12: Avoid Striding and Slicing in a Single Expression

In addition to basic slicing (see Item 11: "Know How to Slice Sequences"), Python has special syntax for the stride of a slice in the form somelist[start:end:stride]. This lets you take every nth item when slicing a sequence. For example, the stride makes it easy to group by even and odd indexes in a list:

```
x = ['red', 'orange', 'yellow', 'green', 'blue', 'purple']
odds = x[::2]
evens = x[1::2]
print(odds)
print(evens)

>>>
['red', 'yellow', 'blue']
['orange', 'green', 'purple']
```

The problem is that the stride syntax often causes unexpected behavior that can introduce bugs. For example, a common Python trick for reversing a byte string is to slice the string with a stride of -1:

```
x = b'mongoose'
y = x[::-1]
print(y)
```

```
>>>
b'esoognom'
```

This also works correctly for Unicode strings (see Item 3: "Know the Differences Between bytes and str"):

```
x = '寿司'
y = x[::-1]
print(y)
```

```
>>>
司寿
```

But it will break when Unicode data is encoded as a UTF-8 byte string:

```
w = '寿司'
x = w.encode('utf-8')
y = x[::-1]
z = y.decode('utf-8')
```

```
>>>
Traceback ...
UnicodeDecodeError: 'utf-8' codec can't decode byte 0xb8 in
position 0: invalid start byte
```

Are negative strides besides -1 useful? Consider the following examples:

```
x = ['a', 'b', 'c', 'd', 'e', 'f', 'g', 'h']
x[::2]   # ['a', 'c', 'e', 'g']
x[::-2]  # ['h', 'f', 'd', 'b']
```

Here, ::2 means "Select every second item starting at the beginning." Trickier, ::-2 means "Select every second item starting at the end and moving backward."

What do you think 2::2 means? What about -2::-2 vs. -2:2:-2 vs. 2:2:-2?

```
x[2::2]    # ['c', 'e', 'g']
x[-2::-2]  # ['g', 'e', 'c', 'a']
x[-2:2:-2] # ['g', 'e']
x[2:2:-2]  # []
```

The point is that the stride part of the slicing syntax can be extremely confusing. Having three numbers within the brackets is hard enough to read because of its density. Then, it's not obvious when the start and end indexes come into effect relative to the stride value, especially when the stride is negative.

To prevent problems, I suggest you avoid using a stride along with start and end indexes. If you must use a stride, prefer making it a positive value and omit start and end indexes. If you must use a stride with start or end indexes, consider using one assignment for striding and another for slicing:

```
y = x[::2]    # ['a', 'c', 'e', 'g']
z = y[1:-1]   # ['c', 'e']
```

Striding and then slicing creates an extra shallow copy of the data. The first operation should try to reduce the size of the resulting slice by as much as possible. If your program can't afford the time or memory required for two steps, consider using the itertools built-in module's islice method (see Item 36: "Consider itertools for Working with Iterators and Generators"), which is clearer to read and doesn't permit negative values for start, end, or stride.

Things to Remember

✦ Specifying start, end, and stride in a slice can be extremely confusing.

✦ Prefer using positive stride values in slices without start or end indexes. Avoid negative stride values if possible.

✦ Avoid using start, end, and stride together in a single slice. If you need all three parameters, consider doing two assignments (one to stride and another to slice) or using islice from the itertools built-in module.

Item 13: Prefer Catch-All Unpacking Over Slicing

One limitation of basic unpacking (see Item 6: "Prefer Multiple Assignment Unpacking Over Indexing") is that you must know the length of the sequences you're unpacking in advance. For example, here I have a list of the ages of cars that are being traded in at a dealership. When I try to take the first two items of the list with basic unpacking, an exception is raised at runtime:

```
car_ages = [0, 9, 4, 8, 7, 20, 19, 1, 6, 15]
car_ages_descending = sorted(car_ages, reverse=True)
oldest, second_oldest = car_ages_descending
```

```
>>>
Traceback ...
ValueError: too many values to unpack (expected 2)
```

Newcomers to Python often rely on indexing and slicing (see Item 11: "Know How to Slice Sequences") for this situation. For example, here I extract the oldest, second oldest, and other car ages from a list of at least two items:

```
oldest = car_ages_descending[0]
second_oldest = car_ages_descending[1]
others = car_ages_descending[2:]
print(oldest, second_oldest, others)
```

```
>>>
20 19 [15, 9, 8, 7, 6, 4, 1, 0]
```

This works, but all of the indexing and slicing is visually noisy. In practice, it's also error prone to divide the members of a sequence into various subsets this way because you're much more likely to make off-by-one errors; for example, you might change boundaries on one line and forget to update the others.

To better handle this situation, Python also supports catch-all unpacking through a *starred expression*. This syntax allows one part of the unpacking assignment to receive all values that didn't match any other part of the unpacking pattern. Here, I use a starred expression to achieve the same result as above without indexing or slicing:

```
oldest, second_oldest, *others = car_ages_descending
print(oldest, second_oldest, others)
```

```
>>>
20 19 [15, 9, 8, 7, 6, 4, 1, 0]
```

This code is shorter, easier to read, and no longer has the error-prone brittleness of boundary indexes that must be kept in sync between lines.

A starred expression may appear in any position, so you can get the benefits of catch-all unpacking anytime you need to extract one slice:

```
oldest, *others, youngest = car_ages_descending
print(oldest, youngest, others)

*others, second_youngest, youngest = car_ages_descending
print(youngest, second_youngest, others)
```

```
>>>
20 0 [19, 15, 9, 8, 7, 6, 4, 1]
0 1 [20, 19, 15, 9, 8, 7, 6, 4]
```

However, to unpack assignments that contain a starred expression, you must have at least one required part, or else you'll get a SyntaxError. You can't use a catch-all expression on its own:

```
*others = car_ages_descending
```

```
>>>
Traceback ...
SyntaxError: starred assignment target must be in a list or
➥tuple
```

You also can't use multiple catch-all expressions in a single-level unpacking pattern:

```
first, *middle, *second_middle, last = [1, 2, 3, 4]
```

```
>>>
Traceback ...
SyntaxError: two starred expressions in assignment
```

But it is possible to use multiple starred expressions in an unpacking assignment statement, as long as they're catch-alls for different parts of the multilevel structure being unpacked. I don't recommend doing the following (see Item 19: "Never Unpack More Than Three Variables When Functions Return Multiple Values" for related guidance), but understanding it should help you develop an intuition for how starred expressions can be used in unpacking assignments:

```
car_inventory = {
    'Downtown': ('Silver Shadow', 'Pinto', 'DMC'),
    'Airport': ('Skyline', 'Viper', 'Gremlin', 'Nova'),
}
```

```
((loc1, (best1, *rest1)),
 (loc2, (best2, *rest2))) = car_inventory.items()
print(f'Best at {loc1} is {best1}, {len(rest1)} others')
print(f'Best at {loc2} is {best2}, {len(rest2)} others')
```

```
>>>
Best at Downtown is Silver Shadow, 2 others
Best at Airport is Skyline, 3 others
```

Starred expressions become list instances in all cases. If there are no leftover items from the sequence being unpacked, the catch-all part will be an empty list. This is especially useful when you're processing a sequence that you know in advance has at least N elements:

```
short_list = [1, 2]
first, second, *rest = short_list
print(first, second, rest)
```

```
>>>
1 2 []
```

You can also unpack arbitrary iterators with the unpacking syntax. This isn't worth much with a basic multiple-assignment statement. For example, here I unpack the values from iterating over a range of length 2. This doesn't seem useful because it would be easier to just assign to a static list that matches the unpacking pattern (e.g., [1, 2]):

```
it = iter(range(1, 3))
first, second = it
print(f'{first} and {second}')
```

```
>>>
1 and 2
```

But with the addition of starred expressions, the value of unpacking iterators becomes clear. For example, here I have a generator that yields the rows of a CSV file containing all car orders from the dealership this week:

```
def generate_csv():
    yield ('Date', 'Make' , 'Model', 'Year', 'Price')
    ...
```

Processing the results of this generator using indexes and slices is fine, but it requires multiple lines and is visually noisy:

```
all_csv_rows = list(generate_csv())
header = all_csv_rows[0]
rows = all_csv_rows[1:]
print('CSV Header:', header)
print('Row count: ', len(rows))
```

```
>>>
CSV Header: ('Date', 'Make', 'Model', 'Year', 'Price')
Row count:  200
```

Unpacking with a starred expression makes it easy to process the first row—the header—separately from the rest of the iterator's contents. This is much clearer:

```
it = generate_csv()
header, *rows = it
print('CSV Header:', header)
print('Row count: ', len(rows))
```

```
>>>
CSV Header: ('Date', 'Make', 'Model', 'Year', 'Price')
Row count:  200
```

Keep in mind, however, that because a starred expression is always turned into a list, unpacking an iterator also risks the potential of using up all of the memory on your computer and causing your program to crash. So you should only use catch-all unpacking on iterators when you have good reason to believe that the result data will all fit in memory (see Item 31: "Be Defensive When Iterating Over Arguments" for another approach).

Things to Remember

✦ Unpacking assignments may use a starred expression to catch all values that weren't assigned to the other parts of the unpacking pattern into a list.

✦ Starred expressions may appear in any position, and they will always become a list containing the zero or more values they receive.

✦ When dividing a list into non-overlapping pieces, catch-all unpacking is much less error prone than slicing and indexing.

Item 14: Sort by Complex Criteria Using the key Parameter

The list built-in type provides a sort method for ordering the items in a list instance based on a variety of criteria. By default, sort will order a list's contents by the natural ascending order of the items. For example, here I sort a list of integers from smallest to largest:

```
numbers = [93, 86, 11, 68, 70]
numbers.sort()
print(numbers)
```

```
>>>
[11, 68, 70, 86, 93]
```

The sort method works for nearly all built-in types (strings, floats, etc.) that have a natural ordering to them. What does sort do with objects? For example, here I define a class—including a __repr__ method so instances are printable; see Item 75: "Use repr Strings for Debugging Output"—to represent various tools you may need to use on a construction site:

```
class Tool:
    def __init__(self, name, weight):
        self.name = name
        self.weight = weight

    def __repr__(self):
        return f'Tool({self.name!r}, {self.weight})'
```

```
tools = [
    Tool('level', 3.5),
    Tool('hammer', 1.25),
    Tool('screwdriver', 0.5),
    Tool('chisel', 0.25),
]
```

Sorting objects of this type doesn't work because the sort method tries to call comparison special methods that aren't defined by the class:

```
tools.sort()
```

```
>>>
Traceback ...
TypeError: '<' not supported between instances of 'Tool' and
'Tool'
```

If your class should have a natural ordering like integers do, then you can define the necessary special methods (see Item 73: "Know How to Use heapq for Priority Queues" for an example) to make sort work without extra parameters. But the more common case is that your objects may need to support multiple orderings, in which case defining a natural ordering really doesn't make sense.

Often there's an attribute on the object that you'd like to use for sorting. To support this use case, the sort method accepts a key parameter that's expected to be a function. The key function is passed a single argument, which is an item from the list that is being sorted. The return value of the key function should be a comparable value (i.e., with a natural ordering) to use in place of an item for sorting purposes.

Here, I use the lambda keyword to define a function for the key parameter that enables me to sort the list of Tool objects alphabetically by their name:

```
print('Unsorted:', repr(tools))
tools.sort(key=lambda x: x.name)
print('\nSorted: ', tools)
```

```
>>>
Unsorted: [Tool('level',      3.5),
           Tool('hammer',     1.25),
           Tool('screwdriver', 0.5),
           Tool('chisel',     0.25)]
```

```
Sorted:    [Tool('chisel',      0.25),
            Tool('hammer',      1.25),
            Tool('level',       3.5),
            Tool('screwdriver', 0.5)]
```

I can just as easily define another lambda function to sort by weight
and pass it as the key parameter to the sort method:

```
tools.sort(key=lambda x: x.weight)
print('By weight:', tools)
```

```
>>>
By weight: [Tool('chisel',      0.25),
            Tool('screwdriver', 0.5),
            Tool('hammer',      1.25),
            Tool('level',       3.5)]
```

Within the lambda function passed as the key parameter you can
access attributes of items as I've done here, index into items (for
sequences, tuples, and dictionaries), or use any other valid expression.

For basic types like strings, you may even want to use the key function to do transformations on the values before sorting. For example,
here I apply the lower method to each item in a list of place names to
ensure that they're in alphabetical order, ignoring any capitalization
(since in the natural lexical ordering of strings, capital letters come
before lowercase letters):

```
places = ['home', 'work', 'New York', 'Paris']
places.sort()
print('Case sensitive:  ', places)
places.sort(key=lambda x: x.lower())
print('Case insensitive:', places)
```

```
>>>
Case sensitive:   ['New York', 'Paris',    'home',    'work']
Case insensitive: ['home',     'New York', 'Paris', 'work']
```

Sometimes you may need to use multiple criteria for sorting. For
example, say that I have a list of power tools and I want to sort them
first by weight and then by name. How can I accomplish this?

```
power_tools = [
    Tool('drill', 4),
    Tool('circular saw', 5),
    Tool('jackhammer', 40),
    Tool('sander', 4),
]
```

The simplest solution in Python is to use the tuple type. Tuples are immutable sequences of arbitrary Python values. Tuples are comparable by default and have a natural ordering, meaning that they implement all of the special methods, such as __lt__, that are required by the sort method. Tuples implement these special method comparators by iterating over each position in the tuple and comparing the corresponding values one index at a time. Here, I show how this works when one tool is heavier than another:

```
saw = (5, 'circular saw')
jackhammer = (40, 'jackhammer')
assert not (jackhammer < saw)  # Matches expectations
```

If the first position in the tuples being compared are equal—weight in this case—then the tuple comparison will move on to the second position, and so on:

```
drill = (4, 'drill')
sander = (4, 'sander')
assert drill[0] == sander[0]  # Same weight
assert drill[1] < sander[1]   # Alphabetically less
assert drill < sander         # Thus, drill comes first
```

You can take advantage of this tuple comparison behavior in order to sort the list of power tools first by weight and then by name. Here, I define a key function that returns a tuple containing the two attributes that I want to sort on in order of priority:

```
power_tools.sort(key=lambda x: (x.weight, x.name))
print(power_tools)

>>>
[Tool('drill',        4),
 Tool('sander',       4),
 Tool('circular saw', 5),
 Tool('jackhammer',   40)]
```

One limitation of having the key function return a tuple is that the direction of sorting for all criteria must be the same (either all in ascending order, or all in descending order). If I provide the reverse parameter to the sort method, it will affect both criteria in the tuple the same way (note how 'sander' now comes before 'drill' instead of after):

```
power_tools.sort(key=lambda x: (x.weight, x.name),
                 reverse=True)  # Makes all criteria descending
print(power_tools)
```

```
>>>
[Tool('jackhammer',    40),
 Tool('circular saw', 5),
 Tool('sander',        4),
 Tool('drill',         4)]
```

For numerical values it's possible to mix sorting directions by using the unary minus operator in the key function. This negates one of the values in the returned tuple, effectively reversing its sort order while leaving the others intact. Here, I use this approach to sort by weight descending, and then by name ascending (note how 'sander' now comes after 'drill' instead of before):

```
power_tools.sort(key=lambda x: (-x.weight, x.name))
print(power_tools)
```

```
>>>
[Tool('jackhammer',    40),
 Tool('circular saw', 5),
 Tool('drill',         4),
 Tool('sander',        4)]
```

Unfortunately, unary negation isn't possible for all types. Here, I try to achieve the same outcome by using the reverse argument to sort by weight descending and then negating name to put it in ascending order:

```
power_tools.sort(key=lambda x: (x.weight, -x.name),
                 reverse=True)
```

```
>>>
Traceback ...
TypeError: bad operand type for unary -: 'str'
```

For situations like this, Python provides a *stable* sorting algorithm. The sort method of the list type will preserve the order of the input list when the key function returns values that are equal to each other. This means that I can call sort multiple times on the same list to combine different criteria together. Here, I produce the same sort ordering of weight descending and name ascending as I did above but by using two separate calls to sort:

```
power_tools.sort(key=lambda x: x.name)    # Name ascending

power_tools.sort(key=lambda x: x.weight, # Weight descending
                 reverse=True)

print(power_tools)
```

```
>>>
[Tool('jackhammer',    40),
 Tool('circular saw', 5),
 Tool('drill',         4),
 Tool('sander',        4)]
```

To understand why this works, note how the first call to sort puts the names in alphabetical order:

```
power_tools.sort(key=lambda x: x.name)
print(power_tools)
```

```
>>>
[Tool('circular saw', 5),
 Tool('drill',         4),
 Tool('jackhammer',    40),
 Tool('sander',        4)]
```

When the second sort call by weight descending is made, it sees that both 'sander' and 'drill' have a weight of 4. This causes the sort method to put both items into the final result list in the same order that they appeared in the original list, thus preserving their relative ordering by name ascending:

```
power_tools.sort(key=lambda x: x.weight,
                 reverse=True)
print(power_tools)
```

```
>>>
[Tool('jackhammer',    40),
 Tool('circular saw', 5),
 Tool('drill',         4),
 Tool('sander',        4)]
```

This same approach can be used to combine as many different types of sorting criteria as you'd like in any direction, respectively. You just need to make sure that you execute the sorts in the opposite sequence of what you want the final list to contain. In this example, I wanted the sort order to be by weight descending and then by name ascending, so I had to do the name sort first, followed by the weight sort.

That said, the approach of having the key function return a tuple, and using unary negation to mix sort orders, is simpler to read and requires less code. I recommend only using multiple calls to sort if it's absolutely necessary.

Things to Remember

+ The sort method of the list type can be used to rearrange a list's contents by the natural ordering of built-in types like strings, integers, tuples, and so on.

+ The sort method doesn't work for objects unless they define a natural ordering using special methods, which is uncommon.

+ The key parameter of the sort method can be used to supply a helper function that returns the value to use for sorting in place of each item from the list.

+ Returning a tuple from the key function allows you to combine multiple sorting criteria together. The unary minus operator can be used to reverse individual sort orders for types that allow it.

+ For types that can't be negated, you can combine many sorting criteria together by calling the sort method multiple times using different key functions and reverse values, in the order of lowest rank sort call to highest rank sort call.

Item 15: Be Cautious When Relying on dict Insertion Ordering

In Python 3.5 and before, iterating over a dict would return keys in arbitrary order. The order of iteration would not match the order in which the items were inserted. For example, here I create a dictionary mapping animal names to their corresponding baby names and then print it out (see Item 75: "Use repr Strings for Debugging Output" for how this works):

```
# Python 3.5
baby_names = {
    'cat': 'kitten',
    'dog': 'puppy',
}
print(baby_names)

>>>
{'dog': 'puppy', 'cat': 'kitten'}
```

When I created the dictionary the keys were in the order 'cat', 'dog', but when I printed it the keys were in the reverse order 'dog', 'cat'. This behavior is surprising, makes it harder to reproduce test cases, increases the difficulty of debugging, and is especially confusing to newcomers to Python.

This happened because the dictionary type previously implemented its hash table algorithm with a combination of the hash built-in function and a random seed that was assigned when the Python interpreter started. Together, these behaviors caused dictionary orderings to not match insertion order and to randomly shuffle between program executions.

Starting with Python 3.6, and officially part of the Python specification in version 3.7, dictionaries will preserve insertion order. Now, this code will always print the dictionary in the same way it was originally created by the programmer:

```
baby_names = {
    'cat': 'kitten',
    'dog': 'puppy',
}
print(baby_names)

>>>
{'cat': 'kitten', 'dog': 'puppy'}
```

With Python 3.5 and earlier, all methods provided by dict that relied on iteration order, including keys, values, items, and popitem, would similarly demonstrate this random-looking behavior:

```
# Python 3.5
print(list(baby_names.keys()))
print(list(baby_names.values()))
print(list(baby_names.items()))
print(baby_names.popitem())  # Randomly chooses an item

>>>
['dog', 'cat']
['puppy', 'kitten']
[('dog', 'puppy'), ('cat', 'kitten')]
('dog', 'puppy')
```

These methods now provide consistent insertion ordering that you can rely on when you write your programs:

```
print(list(baby_names.keys()))
print(list(baby_names.values()))
print(list(baby_names.items()))
print(baby_names.popitem())  # Last item inserted

>>>
['cat', 'dog']
['kitten', 'puppy']
[('cat', 'kitten'), ('dog', 'puppy')]
('dog', 'puppy')
```

There are many repercussions of this change on other Python features that are dependent on the dict type and its specific implementation.

Keyword arguments to functions—including the **kwargs catch-all parameter; see Item 23: "Provide Optional Behavior with Keyword Arguments"—previously would come through in seemingly random order, which can make it harder to debug function calls:

```
# Python 3.5
def my_func(**kwargs):
    for key, value in kwargs.items():
        print('%s = %s' % (key, value))

my_func(goose='gosling', kangaroo='joey')

>>>
kangaroo = joey
goose = gosling
```

Now, the order of keyword arguments is always preserved to match how the programmer originally called the function:

```
def my_func(**kwargs):
    for key, value in kwargs.items():
        print(f'{key} = {value}')

my_func(goose='gosling', kangaroo='joey')

>>>
goose = gosling
kangaroo = joey
```

Classes also use the dict type for their instance dictionaries. In previous versions of Python, object fields would show the randomizing behavior:

```
# Python 3.5
class MyClass:
    def __init__(self):
        self.alligator = 'hatchling'
        self.elephant = 'calf'

a = MyClass()
for key, value in a.__dict__.items():
    print('%s = %s' % (key, value))

>>>
elephant = calf
alligator = hatchling
```

Again, you can now assume that the order of assignment for these instance fields will be reflected in __dict__:

```
class MyClass:
    def __init__(self):
        self.alligator = 'hatchling'
        self.elephant = 'calf'

a = MyClass()
for key, value in a.__dict__.items():
    print(f'{key} = {value}')

>>>
alligator = hatchling
elephant = calf
```

The way that dictionaries preserve insertion ordering is now part of the Python language specification. For the language features above, you can rely on this behavior and even make it part of the APIs you design for your classes and functions.

Note

For a long time the collections built-in module has had an OrderedDict class that preserves insertion ordering. Although this class's behavior is similar to that of the standard dict type (since Python 3.7), the performance characteristics of OrderedDict are quite different. If you need to handle a high rate of key insertions and popitem calls (e.g., to implement a least-recently-used cache), OrderedDict may be a better fit than the standard Python dict type (see Item 70: "Profile Before Optimizing" on how to make sure you need this).

However, you shouldn't always assume that insertion ordering behavior will be present when you're handling dictionaries. Python makes it easy for programmers to define their own custom container types that emulate the standard *protocols* matching list, dict, and other types (see Item 43: "Inherit from collections.abc for Custom Container Types"). Python is not statically typed, so most code relies on *duck typing*—where an object's behavior is its de facto type—instead of rigid class hierarchies. This can result in surprising gotchas.

For example, say that I'm writing a program to show the results of a contest for the cutest baby animal. Here, I start with a dictionary containing the total vote count for each one:

```
votes = {
    'otter': 1281,
    'polar bear': 587,
    'fox': 863,
}
```

I define a function to process this voting data and save the rank of each animal name into a provided empty dictionary. In this case, the dictionary could be the data model that powers a UI element:

```
def populate_ranks(votes, ranks):
    names = list(votes.keys())
    names.sort(key=votes.get, reverse=True)
    for i, name in enumerate(names, 1):
        ranks[name] = i
```

I also need a function that will tell me which animal won the contest. This function works by assuming that populate_ranks will assign the contents of the ranks dictionary in ascending order, meaning that the first key must be the winner:

```
def get_winner(ranks):
    return next(iter(ranks))
```

Here, I can confirm that these functions work as designed and deliver the result that I expected:

```
ranks = {}
populate_ranks(votes, ranks)
print(ranks)
winner = get_winner(ranks)
print(winner)

>>>
{'otter': 1, 'fox': 2, 'polar bear': 3}
otter
```

Now, imagine that the requirements of this program have changed. The UI element that shows the results should be in alphabetical order instead of rank order. To accomplish this, I can use the collections.abc built-in module to define a new dictionary-like class that iterates its contents in alphabetical order:

```
from collections.abc import MutableMapping

class SortedDict(MutableMapping):
    def __init__(self):
        self.data = {}

    def __getitem__(self, key):
        return self.data[key]

    def __setitem__(self, key, value):
        self.data[key] = value
```

```
def __delitem__(self, key):
    del self.data[key]

def __iter__(self):
    keys = list(self.data.keys())
    keys.sort()
    for key in keys:
        yield key

def __len__(self):
    return len(self.data)
```

I can use a SortedDict instance in place of a standard dict with the functions from before and no errors will be raised since this class conforms to the protocol of a standard dictionary. However, the result is incorrect:

```
sorted_ranks = SortedDict()
populate_ranks(votes, sorted_ranks)
print(sorted_ranks.data)
winner = get_winner(sorted_ranks)
print(winner)

>>>
{'otter': 1, 'fox': 2, 'polar bear': 3}
fox
```

The problem here is that the implementation of get_winner assumes that the dictionary's iteration is in insertion order to match populate_ranks. This code is using SortedDict instead of dict, so that assumption is no longer true. Thus, the value returned for the winner is 'fox', which is alphabetically first.

There are three ways to mitigate this problem. First, I can reimplement the get_winner function to no longer assume that the ranks dictionary has a specific iteration order. This is the most conservative and robust solution:

```
def get_winner(ranks):
    for name, rank in ranks.items():
        if rank == 1:
            return name

winner = get_winner(sorted_ranks)
print(winner)

>>>
otter
```

The second approach is to add an explicit check to the top of the function to ensure that the type of ranks matches my expectations, and to raise an exception if not. This solution likely has better runtime performance than the more conservative approach:

```
def get_winner(ranks):
    if not isinstance(ranks, dict):
        raise TypeError('must provide a dict instance')
    return next(iter(ranks))

get_winner(sorted_ranks)

>>>
Traceback ...
TypeError: must provide a dict instance
```

The third alternative is to use type annotations to enforce that the value passed to get_winner is a dict instance and not a MutableMapping with dictionary-like behavior (see Item 90: "Consider Static Analysis via typing to Obviate Bugs"). Here, I run the mypy tool in strict mode on an annotated version of the code above:

```
from typing import Dict, MutableMapping

def populate_ranks(votes: Dict[str, int],
                   ranks: Dict[str, int]) -> None:
    names = list(votes.keys())
    names.sort(key=votes.get, reverse=True)
    for i, name in enumerate(names, 1):
        ranks[name] = i

def get_winner(ranks: Dict[str, int]) -> str:
    return next(iter(ranks))

class SortedDict(MutableMapping[str, int]):
    ...

votes = {
    'otter': 1281,
    'polar bear': 587,
    'fox': 863,
}

sorted_ranks = SortedDict()
populate_ranks(votes, sorted_ranks)
print(sorted_ranks.data)
winner = get_winner(sorted_ranks)
print(winner)
```

```
$ python3 -m mypy --strict example.py
.../example.py:48: error: Argument 2 to "populate_ranks" has
➥incompatible type "SortedDict"; expected "Dict[str, int]"
.../example.py:50: error: Argument 1 to "get_winner" has
➥incompatible type "SortedDict"; expected "Dict[str, int]"
```

This correctly detects the mismatch between the dict and MutableMapping types and flags the incorrect usage as an error. This solution provides the best mix of static type safety and runtime performance.

Things to Remember

✦ Since Python 3.7, you can rely on the fact that iterating a dict instance's contents will occur in the same order in which the keys were initially added.

✦ Python makes it easy to define objects that act like dictionaries but that aren't dict instances. For these types, you can't assume that insertion ordering will be preserved.

✦ There are three ways to be careful about dictionary-like classes: Write code that doesn't rely on insertion ordering, explicitly check for the dict type at runtime, or require dict values using type annotations and static analysis.

Item 16: Prefer get Over in and KeyError to Handle Missing Dictionary Keys

The three fundamental operations for interacting with dictionaries are accessing, assigning, and deleting keys and their associated values. The contents of dictionaries are dynamic, and thus it's entirely possible—even likely—that when you try to access or delete a key, it won't already be present.

For example, say that I'm trying to determine people's favorite type of bread to devise the menu for a sandwich shop. Here, I define a dictionary of counters with the current votes for each style:

```
counters = {
    'pumpernickel': 2,
    'sourdough': 1,
}
```

To increment the counter for a new vote, I need to see if the key exists, insert the key with a default counter value of zero if it's missing, and then increment the counter's value. This requires accessing the key two times and assigning it once. Here, I accomplish this task using an if statement with an in expression that returns True when the key is present:

```
key = 'wheat'

if key in counters:
    count = counters[key]
else:
    count = 0

counters[key] = count + 1
```

Another way to accomplish the same behavior is by relying on how dictionaries raise a KeyError exception when you try to get the value for a key that doesn't exist. This approach is more efficient because it requires only one access and one assignment:

```
try:
    count = counters[key]
except KeyError:
    count = 0

counters[key] = count + 1
```

This flow of fetching a key that exists or returning a default value is so common that the dict built-in type provides the get method to accomplish this task. The second parameter to get is the default value to return in the case that the key—the first parameter—isn't present. This also requires only one access and one assignment, but it's much shorter than the KeyError example:

```
count = counters.get(key, 0)
counters[key] = count + 1
```

It's possible to shorten the in expression and KeyError approaches in various ways, but all of these alternatives suffer from requiring code duplication for the assignments, which makes them less readable and worth avoiding:

```
if key not in counters:
    counters[key] = 0
counters[key] += 1

if key in counters:
    counters[key] += 1
else:
    counters[key] = 1
```

```
try:
    counters[key] += 1
except KeyError:
    counters[key] = 1
```

Thus, for a dictionary with simple types, using the get method is the shortest and clearest option.

> Note
>
> If you're maintaining dictionaries of counters like this, it's worth considering the Counter class from the collections built-in module, which provides most of the facilities you are likely to need.

What if the values of the dictionary are a more complex type, like a list? For example, say that instead of only counting votes, I also want to know who voted for each type of bread. Here, I do this by associating a list of names with each key:

```
votes = {
    'baguette': ['Bob', 'Alice'],
    'ciabatta': ['Coco', 'Deb'],
}
key = 'brioche'
who = 'Elmer'

if key in votes:
    names = votes[key]
else:
    votes[key] = names = []

names.append(who)
print(votes)

>>>
{'baguette': ['Bob', 'Alice'],
 'ciabatta': ['Coco', 'Deb'],
 'brioche': ['Elmer']}
```

Relying on the in expression requires two accesses if the key is present, or one access and one assignment if the key is missing. This example is different from the counters example above because the value for each key can be assigned blindly to the default value of an empty list if the key doesn't already exist. The triple assignment statement (votes[key] = names = []) populates the key in one line instead of two. Once the default value has been inserted into the dictionary, I don't need to assign it again because the list is modified by reference in the later call to append.

It's also possible to rely on the KeyError exception being raised when the dictionary value is a list. This approach requires one key access if the key is present, or one key access and one assignment if it's missing, which makes it more efficient than the in condition:

```
try:
    names = votes[key]
except KeyError:
    votes[key] = names = []
```

```
names.append(who)
```

Similarly, you can use the get method to fetch a list value when the key is present, or do one fetch and one assignment if the key isn't present:

```
names = votes.get(key)
if names is None:
    votes[key] = names = []
```

```
names.append(who)
```

The approach that involves using get to fetch list values can further be shortened by one line if you use an assignment expression (introduced in Python 3.8; see Item 10: "Prevent Repetition with Assignment Expressions") in the if statement, which improves readability:

```
if (names := votes.get(key)) is None:
    votes[key] = names = []
```

```
names.append(who)
```

The dict type also provides the setdefault method to help shorten this pattern even further. setdefault tries to fetch the value of a key in the dictionary. If the key isn't present, the method assigns that key to the default value provided. And then the method returns the value for that key: either the originally present value or the newly inserted default value. Here, I use setdefault to implement the same logic as in the get example above:

```
names = votes.setdefault(key, [])
names.append(who)
```

This works as expected, and it is shorter than using get with an assignment expression. However, the readability of this approach isn't ideal. The method name setdefault doesn't make its purpose immediately obvious. Why is it set when what it's doing is getting a value? Why not call it get_or_set? I'm arguing about the color of the bike shed here, but the point is that if you were a new reader of the code and not completely familiar with Python, you might have trouble understanding what this code is trying to accomplish because setdefault isn't self-explanatory.

There's also one important gotcha: The default value passed to setdefault is assigned directly into the dictionary when the key is missing instead of being copied. Here, I demonstrate the effect of this when the value is a list:

```
data = {}
key = 'foo'
value = []
data.setdefault(key, value)
print('Before:', data)
value.append('hello')
print('After: ', data)

>>>
Before: {'foo': []}
After:  {'foo': ['hello']}
```

This means that I need to make sure that I'm always constructing a new default value for each key I access with setdefault. This leads to a significant performance overhead in this example because I have to allocate a list instance for each call. If I reuse an object for the default value—which I might try to do to increase efficiency or readability—I might introduce strange behavior and bugs (see Item 24: "Use None and Docstrings to Specify Dynamic Default Arguments" for another example of this problem).

Going back to the earlier example that used counters for dictionary values instead of lists of who voted: Why not also use the setdefault method in that case? Here, I reimplement the same example using this approach:

```
count = counters.setdefault(key, 0)
counters[key] = count + 1
```

The problem here is that the call to setdefault is superfluous. You always need to assign the key in the dictionary to a new value after you increment the counter, so the extra assignment done by setdefault is unnecessary. The earlier approach of using get for counter updates requires only one access and one assignment, whereas using setdefault requires one access and two assignments.

There are only a few circumstances in which using setdefault is the shortest way to handle missing dictionary keys, such as when the default values are cheap to construct, mutable, and there's no potential for raising exceptions (e.g., list instances). In these very specific cases, it may seem worth accepting the confusing method name setdefault instead of having to write more characters and lines to use get. However, often what you really should do in these situations is to use defaultdict instead (see Item 17: "Prefer defaultdict Over setdefault to Handle Missing Items in Internal State").

Things to Remember

+ There are four common ways to detect and handle missing keys in dictionaries: using in expressions, KeyError exceptions, the get method, and the setdefault method.

+ The get method is best for dictionaries that contain basic types like counters, and it is preferable along with assignment expressions when creating dictionary values has a high cost or may raise exceptions.

+ When the setdefault method of dict seems like the best fit for your problem, you should consider using defaultdict instead.

Item 17: Prefer defaultdict Over setdefault to Handle Missing Items in Internal State

When working with a dictionary that you didn't create, there are a variety of ways to handle missing keys (see Item 16: "Prefer get Over in and KeyError to Handle Missing Dictionary Keys"). Although using the get method is a better approach than using in expressions and KeyError exceptions, for some use cases setdefault appears to be the shortest option.

For example, say that I want to keep track of the cities I've visited in countries around the world. Here, I do this by using a dictionary that maps country names to a set instance containing corresponding city names:

```
visits = {
    'Mexico': {'Tulum', 'Puerto Vallarta'},
    'Japan': {'Hakone'},
}
```

I can use the setdefault method to add new cities to the sets, whether the country name is already present in the dictionary or not. This approach is much shorter than achieving the same behavior with the get method and an assignment expression (which is available as of Python 3.8):

```python
visits.setdefault('France', set()).add('Arles')   # Short

if (japan := visits.get('Japan')) is None:        # Long
    visits['Japan'] = japan = set()
japan.add('Kyoto')

print(visits)

>>>
{'Mexico': {'Tulum', 'Puerto Vallarta'},
 'Japan': {'Kyoto', 'Hakone'},
 'France': {'Arles'}}
```

What about the situation when you *do* control creation of the dictionary being accessed? This is generally the case when you're using a dictionary instance to keep track of the internal state of a class, for example. Here, I wrap the example above in a class with helper methods to access the dynamic inner state stored in a dictionary:

```python
class Visits:
    def __init__(self):
        self.data = {}

    def add(self, country, city):
        city_set = self.data.setdefault(country, set())
        city_set.add(city)
```

This new class hides the complexity of calling setdefault correctly, and it provides a nicer interface for the programmer:

```python
visits = Visits()
visits.add('Russia', 'Yekaterinburg')
visits.add('Tanzania', 'Zanzibar')
print(visits.data)

>>>
{'Russia': {'Yekaterinburg'}, 'Tanzania': {'Zanzibar'}}
```

However, the implementation of the Visits.add method still isn't ideal. The setdefault method is still confusingly named, which makes it more difficult for a new reader of the code to immediately understand what's happening. And the implementation isn't efficient because it constructs a new set instance on every call, regardless of whether the given country was already present in the data dictionary.

Luckily, the defaultdict class from the collections built-in module simplifies this common use case by automatically storing a default value when a key doesn't exist. All you have to do is provide a function that will return the default value to use each time a key is missing (an example of Item 38: "Accept Functions Instead of Classes for Simple Interfaces"). Here, I rewrite the Visits class to use defaultdict:

```
from collections import defaultdict

class Visits:
    def __init__(self):
        self.data = defaultdict(set)

    def add(self, country, city):
        self.data[country].add(city)

visits = Visits()
visits.add('England', 'Bath')
visits.add('England', 'London')
print(visits.data)

>>>
defaultdict(<class 'set'>, {'England': {'London', 'Bath'}})
```

Now, the implementation of add is short and simple. The code can assume that accessing any key in the data dictionary will always result in an existing set instance. No superfluous set instances will be allocated, which could be costly if the add method is called a large number of times.

Using defaultdict is much better than using setdefault for this type of situation (see Item 37: "Compose Classes Instead of Nesting Many Levels of Built-in Types" for another example). There are still cases in which defaultdict will fall short of solving your problems, but there are even more tools available in Python to work around those limitations (see Item 18: "Know How to Construct Key-Dependent Default Values with __missing__," Item 43: "Inherit from collections.abc for Custom Container Types," and the collections.Counter built-in class).

Things to Remember

✦ If you're creating a dictionary to manage an arbitrary set of potential keys, then you should prefer using a defaultdict instance from the collections built-in module if it suits your problem.

✦ If a dictionary of arbitrary keys is passed to you, and you don't control its creation, then you should prefer the get method to access its items. However, it's worth considering using the setdefault method for the few situations in which it leads to shorter code.

Item 18: Know How to Construct Key-Dependent Default Values with __missing__

The built-in dict type's setdefault method results in shorter code when handling missing keys in some specific circumstances (see Item 16: "Prefer get Over in and KeyError to Handle Missing Dictionary Keys" for examples). For many of those situations, the better tool for the job is the defaultdict type from the collections built-in module (see Item 17: "Prefer defaultdict Over setdefault to Handle Missing Items in Internal State" for why). However, there are times when neither setdefault nor defaultdict is the right fit.

For example, say that I'm writing a program to manage social network profile pictures on the filesystem. I need a dictionary to map profile picture pathnames to open file handles so I can read and write those images as needed. Here, I do this by using a normal dict instance and checking for the presence of keys using the get method and an assignment expression (introduced in Python 3.8; see Item 10: "Prevent Repetition with Assignment Expressions"):

```
pictures = {}
path = 'profile_1234.png'

if (handle := pictures.get(path)) is None:
    try:
        handle = open(path, 'a+b')
    except OSError:
        print(f'Failed to open path {path}')
        raise
    else:
        pictures[path] = handle

handle.seek(0)
image_data = handle.read()
```

When the file handle already exists in the dictionary, this code makes only a single dictionary access. In the case that the file handle doesn't exist, the dictionary is accessed once by get, and then it is assigned in the else clause of the try/except block. (This approach also works with finally; see Item 65: "Take Advantage of Each Block in try/except/else/finally.") The call to the read method stands clearly separate from the code that calls open and handles exceptions.

Although it's possible to use the in expression or KeyError approaches to implement this same logic, those options require more dictionary accesses and levels of nesting. Given that these other options work, you might also assume that the setdefault method would work, too:

```
try:
    handle = pictures.setdefault(path, open(path, 'a+b'))
except OSError:
    print(f'Failed to open path {path}')
    raise
else:
    handle.seek(0)
    image_data = handle.read()
```

This code has many problems. The open built-in function to create the file handle is always called, even when the path is already present in the dictionary. This results in an additional file handle that may conflict with existing open handles in the same program. Exceptions may be raised by the open call and need to be handled, but it may not be possible to differentiate them from exceptions that may be raised by the setdefault call on the same line (which is possible for other dictionary-like implementations; see Item 43: "Inherit from collections.abc for Custom Container Types").

If you're trying to manage internal state, another assumption you might make is that a defaultdict could be used for keeping track of these profile pictures. Here, I attempt to implement the same logic as before but now using a helper function and the defaultdict class:

```
from collections import defaultdict

def open_picture(profile_path):
    try:
        return open(profile_path, 'a+b')
    except OSError:
        print(f'Failed to open path {profile_path}')
        raise

pictures = defaultdict(open_picture)
handle = pictures[path]
handle.seek(0)
image_data = handle.read()

>>>
Traceback ...
TypeError: open_picture() missing 1 required positional
argument: 'profile_path'
```

The problem is that defaultdict expects that the function passed to its constructor doesn't require any arguments. This means that the helper function that defaultdict calls doesn't know which specific key is being accessed, which eliminates my ability to call open. In this situation, both setdefault and defaultdict fall short of what I need.

Fortunately, this situation is common enough that Python has another built-in solution. You can subclass the dict type and implement the __missing__ special method to add custom logic for handling missing keys. Here, I do this by defining a new class that takes advantage of the same open_picture helper method defined above:

```
class Pictures(dict):
    def __missing__(self, key):
        value = open_picture(key)
        self[key] = value
        return value
```

```
pictures = Pictures()
handle = pictures[path]
handle.seek(0)
image_data = handle.read()
```

When the pictures[path] dictionary access finds that the path key isn't present in the dictionary, the __missing__ method is called. This method must create the new default value for the key, insert it into the dictionary, and return it to the caller. Subsequent accesses of the same path will not call __missing__ since the corresponding item is already present (similar to the behavior of __getattr__; see Item 47: "Use __getattr__, __getattribute__, and __setattr__ for Lazy Attributes").

Things to Remember

✦ The setdefault method of dict is a bad fit when creating the default value has high computational cost or may raise exceptions.

✦ The function passed to defaultdict must not require any arguments, which makes it impossible to have the default value depend on the key being accessed.

✦ You can define your own dict subclass with a __missing__ method in order to construct default values that must know which key was being accessed.

Chapter **3** Functions

The first organizational tool programmers use in Python is the *function*. As in other programming languages, functions enable you to break large programs into smaller, simpler pieces with names to represent their intent. They improve readability and make code more approachable. They allow for reuse and refactoring.

Functions in Python have a variety of extra features that make a programmer's life easier. Some are similar to capabilities in other programming languages, but many are unique to Python. These extras can make a function's purpose more obvious. They can eliminate noise and clarify the intention of callers. They can significantly reduce subtle bugs that are difficult to find.

Item 19: Never Unpack More Than Three Variables When Functions Return Multiple Values

One effect of the unpacking syntax (see Item 6: "Prefer Multiple Assignment Unpacking Over Indexing") is that it allows Python functions to seemingly return more than one value. For example, say that I'm trying to determine various statistics for a population of alligators. Given a `list` of lengths, I need to calculate the minimum and maximum lengths in the population. Here, I do this in a single function that appears to return two values:

```python
def get_stats(numbers):
    minimum = min(numbers)
    maximum = max(numbers)
    return minimum, maximum

lengths = [63, 73, 72, 60, 67, 66, 71, 61, 72, 70]

minimum, maximum = get_stats(lengths)  # Two return values

print(f'Min: {minimum}, Max: {maximum}')
```

```
>>>
Min: 60, Max: 73
```

The way this works is that multiple values are returned together in a two-item `tuple`. The calling code then unpacks the returned `tuple` by assigning two variables. Here, I use an even simpler example to show how an unpacking statement and multiple-return function work the same way:

```
first, second = 1, 2
assert first == 1
assert second == 2

def my_function():
    return 1, 2

first, second = my_function()
assert first == 1
assert second == 2
```

Multiple return values can also be received by starred expressions for catch-all unpacking (see Item 13: "Prefer Catch-All Unpacking Over Slicing"). For example, say I need another function that calculates how big each alligator is relative to the population average. This function returns a list of ratios, but I can receive the longest and shortest items individually by using a starred expression for the middle portion of the list:

```
def get_avg_ratio(numbers):
    average = sum(numbers) / len(numbers)
    scaled = [x / average for x in numbers]
    scaled.sort(reverse=True)
    return scaled

longest, *middle, shortest = get_avg_ratio(lengths)

print(f'Longest:  {longest:>4.0%}')
print(f'Shortest: {shortest:>4.0%}')
```

```
>>>
Longest:  108%
Shortest:  89%
```

Now, imagine that the program's requirements change, and I need to also determine the average length, median length, and total population size of the alligators. I can do this by expanding the `get_stats` function to also calculate these statistics and return them in the result `tuple` that is unpacked by the caller:

```
def get_stats(numbers):
    minimum = min(numbers)
    maximum = max(numbers)
    count = len(numbers)
    average = sum(numbers) / count

    sorted_numbers = sorted(numbers)
    middle = count // 2
    if count % 2 == 0:
        lower = sorted_numbers[middle - 1]
        upper = sorted_numbers[middle]
        median = (lower + upper) / 2
    else:
        median = sorted_numbers[middle]

    return minimum, maximum, average, median, count

minimum, maximum, average, median, count = get_stats(lengths)

print(f'Min: {minimum}, Max: {maximum}')
print(f'Average: {average}, Median: {median}, Count {count}')

>>>
Min: 60, Max: 73
Average: 67.5, Median: 68.5, Count 10
```

There are two problems with this code. First, all the return values are numeric, so it is all too easy to reorder them accidentally (e.g., swapping average and median), which can cause bugs that are hard to spot later. Using a large number of return values is extremely error prone:

```
# Correct:
minimum, maximum, average, median, count = get_stats(lengths)

# Oops! Median and average swapped:
minimum, maximum, median, average, count = get_stats(lengths)
```

Second, the line that calls the function and unpacks the values is long, and it likely will need to be wrapped in one of a variety of ways (due to PEP8 style; see Item 2: "Follow the PEP 8 Style Guide"), which hurts readability:

```
minimum, maximum, average, median, count = get_stats(
    lengths)
```

```
minimum, maximum, average, median, count = \
    get_stats(lengths)

(minimum, maximum, average,
 median, count) = get_stats(lengths)

(minimum, maximum, average, median, count
    ) = get_stats(lengths)
```

To avoid these problems, you should never use more than three variables when unpacking the multiple return values from a function. These could be individual values from a three-tuple, two variables and one catch-all starred expression, or anything shorter. If you need to unpack more return values than that, you're better off defining a lightweight class or namedtuple (see Item 37: "Compose Classes Instead of Nesting Many Levels of Built-in Types") and having your function return an instance of that instead.

Things to Remember

✦ You can have functions return multiple values by putting them in a tuple and having the caller take advantage of Python's unpacking syntax.

✦ Multiple return values from a function can also be unpacked by catch-all starred expressions.

✦ Unpacking into four or more variables is error prone and should be avoided; instead, return a small class or namedtuple instance.

Item 20: Prefer Raising Exceptions to Returning None

When writing utility functions, there's a draw for Python programmers to give special meaning to the return value of None. It seems to make sense in some cases. For example, say I want a helper function that divides one number by another. In the case of dividing by zero, returning None seems natural because the result is undefined:

```
def careful_divide(a, b):
    try:
        return a / b
    except ZeroDivisionError:
        return None
```

Code using this function can interpret the return value accordingly:

```
x, y = 1, 0
result = careful_divide(x, y)
if result is None:
    print('Invalid inputs')
```

What happens with the careful_divide function when the numerator is zero? If the denominator is not zero, the function returns zero. The problem is that a zero return value can cause issues when you evaluate the result in a condition like an if statement. You might accidentally look for any False-equivalent value to indicate errors instead of only looking for None (see Item 5: "Write Helper Functions Instead of Complex Expressions" for a similar situation):

```
x, y = 0, 5
result = careful_divide(x, y)
if not result:
    print('Invalid inputs')  # This runs! But shouldn't

>>>
Invalid inputs
```

This misinterpretation of a False-equivalent return value is a common mistake in Python code when None has special meaning. This is why returning None from a function like careful_divide is error prone. There are two ways to reduce the chance of such errors.

The first way is to split the return value into a two-tuple (see Item 19: "Never Unpack More Than Three Variables When Functions Return Multiple Values" for background). The first part of the tuple indicates that the operation was a success or failure. The second part is the actual result that was computed:

```
def careful_divide(a, b):
    try:
        return True, a / b
    except ZeroDivisionError:
        return False, None
```

Callers of this function have to unpack the tuple. That forces them to consider the status part of the tuple instead of just looking at the result of division:

```
success, result = careful_divide(x, y)
if not success:
    print('Invalid inputs')
```

The problem is that callers can easily ignore the first part of the tuple (using the underscore variable name, a Python convention for unused variables). The resulting code doesn't look wrong at first glance, but this can be just as error prone as returning None:

```
_, result = careful_divide(x, y)
if not result:
    print('Invalid inputs')
```

The second, better way to reduce these errors is to never return None for special cases. Instead, raise an Exception up to the caller and have the caller deal with it. Here, I turn a ZeroDivisionError into a ValueError to indicate to the caller that the input values are bad (see Item 87: "Define a Root Exception to Insulate Callers from APIs" on when you should use Exception subclasses):

```
def careful_divide(a, b):
    try:
        return a / b
    except ZeroDivisionError as e:
        raise ValueError('Invalid inputs')
```

The caller no longer requires a condition on the return value of the function. Instead, it can assume that the return value is always valid and use the results immediately in the else block after try (see Item 65: "Take Advantage of Each Block in try/except/else/finally" for details):

```
x, y = 5, 2
try:
    result = careful_divide(x, y)
except ValueError:
    print('Invalid inputs')
else:
    print('Result is %.1f' % result)

>>>
Result is 2.5
```

This approach can be extended to code using type annotations (see Item 90: "Consider Static Analysis via typing to Obviate Bugs" for background). You can specify that a function's return value will always be a float and thus will never be None. However, Python's gradual typing purposefully doesn't provide a way to indicate when exceptions are part of a function's interface (also known as *checked exceptions*). Instead, you have to document the exception-raising behavior and expect callers to rely on that in order to know which Exceptions they should plan to catch (see Item 84: "Write Docstrings for Every Function, Class, and Module").

Pulling it all together, here's what this function should look like when using type annotations and docstrings:

```
def careful_divide(a: float, b: float) -> float:
    """Divides a by b.

    Raises:
        ValueError: When the inputs cannot be divided.
    """
    try:
        return a / b
    except ZeroDivisionError as e:
        raise ValueError('Invalid inputs')
```

Now the inputs, outputs, and exceptional behavior is clear, and the chance of a caller doing the wrong thing is extremely low.

Things to Remember

✦ Functions that return None to indicate special meaning are error prone because None and other values (e.g., zero, the empty string) all evaluate to False in conditional expressions.

✦ Raise exceptions to indicate special situations instead of returning None. Expect the calling code to handle exceptions properly when they're documented.

✦ Type annotations can be used to make it clear that a function will never return the value None, even in special situations.

Item 21: Know How Closures Interact with Variable Scope

Say that I want to sort a list of numbers but prioritize one group of numbers to come first. This pattern is useful when you're rendering a user interface and want important messages or exceptional events to be displayed before everything else.

A common way to do this is to pass a helper function as the key argument to a list's sort method (see Item 14: "Sort by Complex Criteria Using the key Parameter" for details). The helper's return value will be used as the value for sorting each item in the list. The helper can check whether the given item is in the important group and can vary the sorting value accordingly:

```
def sort_priority(values, group):
    def helper(x):
        if x in group:
            return (0, x)
        return (1, x)
    values.sort(key=helper)
```

This function works for simple inputs:

```
numbers = [8, 3, 1, 2, 5, 4, 7, 6]
group = {2, 3, 5, 7}
sort_priority(numbers, group)
print(numbers)

>>>
[2, 3, 5, 7, 1, 4, 6, 8]
```

There are three reasons this function operates as expected:

- Python supports *closures*—that is, functions that refer to variables from the scope in which they were defined. This is why the `helper` function is able to access the `group` argument for `sort_priority`.

- Functions are *first-class* objects in Python, which means you can refer to them directly, assign them to variables, pass them as arguments to other functions, compare them in expressions and `if` statements, and so on. This is how the `sort` method can accept a closure function as the key argument.

- Python has specific rules for comparing sequences (including tuples). It first compares items at index zero; then, if those are equal, it compares items at index one; if they are still equal, it compares items at index two, and so on. This is why the return value from the `helper` closure causes the sort order to have two distinct groups.

It'd be nice if this function returned whether higher-priority items were seen at all so the user interface code can act accordingly. Adding such behavior seems straightforward. There's already a closure function for deciding which group each number is in. Why not also use the closure to flip a flag when high-priority items are seen? Then, the function can return the flag value after it's been modified by the closure.

Here, I try to do that in a seemingly obvious way:

```
def sort_priority2(numbers, group):
    found = False
    def helper(x):
        if x in group:
            found = True   # Seems simple
            return (0, x)
        return (1, x)
    numbers.sort(key=helper)
    return found
```

I can run the function on the same inputs as before:

```
found = sort_priority2(numbers, group)
print('Found:', found)
print(numbers)
```

```
>>>
Found: False
[2, 3, 5, 7, 1, 4, 6, 8]
```

The sorted results are correct, which means items from group were definitely found in numbers. Yet the found result returned by the function is False when it should be True. How could this happen?

When you reference a variable in an expression, the Python interpreter traverses the scope to resolve the reference in this order:

1. The current function's scope.

2. Any enclosing scopes (such as other containing functions).

3. The scope of the module that contains the code (also called the *global scope*).

4. The built-in scope (that contains functions like len and str).

If none of these places has defined a variable with the referenced name, then a NameError exception is raised:

```
foo = does_not_exist * 5
```

```
>>>
Traceback ...
NameError: name 'does_not_exist' is not defined
```

Assigning a value to a variable works differently. If the variable is already defined in the current scope, it will just take on the new value. If the variable doesn't exist in the current scope, Python treats the assignment as a variable definition. Critically, the scope of the newly defined variable is the function that contains the assignment.

This assignment behavior explains the wrong return value of the sort_priority2 function. The found variable is assigned to True in the helper closure. The closure's assignment is treated as a new variable definition within helper, not as an assignment within sort_priority2:

```
def sort_priority2(numbers, group):
    found = False          # Scope: 'sort_priority2'
    def helper(x):
        if x in group:
            found = True   # Scope: 'helper' -- Bad!
            return (0, x)
        return (1, x)
```

```
    numbers.sort(key=helper)
    return found
```

This problem is sometimes called the *scoping bug* because it can be so surprising to newbies. But this behavior is the intended result: It prevents local variables in a function from polluting the containing module. Otherwise, every assignment within a function would put garbage into the global module scope. Not only would that be noise, but the interplay of the resulting global variables could cause obscure bugs.

In Python, there is special syntax for getting data out of a closure. The nonlocal statement is used to indicate that scope traversal should happen upon assignment for a specific variable name. The only limit is that nonlocal won't traverse up to the module-level scope (to avoid polluting globals).

Here, I define the same function again, now using nonlocal:

```
def sort_priority3(numbers, group):
    found = False
    def helper(x):
        nonlocal found  # Added
        if x in group:
            found = True
            return (0, x)
        return (1, x)
    numbers.sort(key=helper)
    return found
```

The nonlocal statement makes it clear when data is being assigned out of a closure and into another scope. It's complementary to the global statement, which indicates that a variable's assignment should go directly into the module scope.

However, much as with the anti-pattern of global variables, I'd caution against using nonlocal for anything beyond simple functions. The side effects of nonlocal can be hard to follow. It's especially hard to understand in long functions where the nonlocal statements and assignments to associated variables are far apart.

When your usage of nonlocal starts getting complicated, it's better to wrap your state in a helper class. Here, I define a class that achieves the same result as the nonlocal approach; it's a little longer but much easier to read (see Item 38: "Accept Functions Instead of Classes for Simple Interfaces" for details on the __call__ special method):

```
class Sorter:
    def __init__(self, group):
```

```
        self.group = group
        self.found = False

    def __call__(self, x):
        if x in self.group:
            self.found = True
            return (0, x)
        return (1, x)

sorter = Sorter(group)
numbers.sort(key=sorter)
assert sorter.found is True
```

Things to Remember

✦ Closure functions can refer to variables from any of the scopes in which they were defined.

✦ By default, closures can't affect enclosing scopes by assigning variables.

✦ Use the nonlocal statement to indicate when a closure can modify a variable in its enclosing scopes.

✦ Avoid using nonlocal statements for anything beyond simple functions.

Item 22: Reduce Visual Noise with Variable Positional Arguments

Accepting a variable number of positional arguments can make a function call clearer and reduce visual noise. (These positional arguments are often called *varargs* for short, or *star args*, in reference to the conventional name for the parameter *args.) For example, say that I want to log some debugging information. With a fixed number of arguments, I would need a function that takes a message and a list of values:

```
def log(message, values):
    if not values:
        print(message)
    else:
        values_str = ', '.join(str(x) for x in values)
        print(f'{message}: {values_str}')

log('My numbers are', [1, 2])
log('Hi there', [])
```

```
>>>
My numbers are: 1, 2
Hi there
```

Having to pass an empty list when I have no values to log is cumbersome and noisy. It'd be better to leave out the second argument entirely. I can do this in Python by prefixing the last positional parameter name with *. The first parameter for the log message is required, whereas any number of subsequent positional arguments are optional. The function body doesn't need to change; only the callers do:

```python
def log(message, *values):  # The only difference
    if not values:
        print(message)
    else:
        values_str = ', '.join(str(x) for x in values)
        print(f'{message}: {values_str}')

log('My numbers are', 1, 2)
log('Hi there')  # Much better
```

```
>>>
My numbers are: 1, 2
Hi there
```

You might notice that this syntax works very similarly to the starred expressions used in unpacking assignment statements (see Item 13: "Prefer Catch-All Unpacking Over Slicing").

If I already have a sequence (like a list) and want to call a variadic function like log, I can do this by using the * operator. This instructs Python to pass items from the sequence as positional arguments to the function:

```python
favorites = [7, 33, 99]
log('Favorite colors', *favorites)
```

```
>>>
Favorite colors: 7, 33, 99
```

There are two problems with accepting a variable number of positional arguments.

The first issue is that these optional positional arguments are always turned into a tuple before they are passed to a function. This means that if the caller of a function uses the * operator on a generator, it will be iterated until it's exhausted (see Item 30: "Consider Generators Instead of Returning Lists" for background). The resulting tuple includes every value from the generator, which could consume a lot of memory and cause the program to crash:

```
def my_generator():
    for i in range(10):
        yield i

def my_func(*args):
    print(args)

it = my_generator()
my_func(*it)

>>>
(0, 1, 2, 3, 4, 5, 6, 7, 8, 9)
```

Functions that accept *args are best for situations where you know the number of inputs in the argument list will be reasonably small. *args is ideal for function calls that pass many literals or variable names together. It's primarily for the convenience of the programmer and the readability of the code.

The second issue with *args is that you can't add new positional arguments to a function in the future without migrating every caller. If you try to add a positional argument in the front of the argument list, existing callers will subtly break if they aren't updated:

```
def log(sequence, message, *values):
    if not values:
        print(f'{sequence} - {message}')
    else:
        values_str = ', '.join(str(x) for x in values)
        print(f'{sequence} - {message}: {values_str}')

log(1, 'Favorites', 7, 33)       # New with *args OK
log(1, 'Hi there')               # New message only OK
log('Favorite numbers', 7, 33)   # Old usage breaks

>>>
1 - Favorites: 7, 33
1 - Hi there
Favorite numbers - 7: 33
```

The problem here is that the third call to log used 7 as the message parameter because a sequence argument wasn't given. Bugs like this are hard to track down because the code still runs without raising exceptions. To avoid this possibility entirely, you should use keyword-only arguments when you want to extend functions that accept *args (see Item 25: "Enforce Clarity with Keyword-Only and Positional-Only Arguments"). To be even more defensive, you could also consider using type annotations (see Item 90: "Consider Static Analysis via typing to Obviate Bugs").

Things to Remember

✦ Functions can accept a variable number of positional arguments by using *args in the def statement.

✦ You can use the items from a sequence as the positional arguments for a function with the * operator.

✦ Using the * operator with a generator may cause a program to run out of memory and crash.

✦ Adding new positional parameters to functions that accept *args can introduce hard-to-detect bugs.

Item 23: Provide Optional Behavior with Keyword Arguments

As in most other programming languages, in Python you may pass arguments by position when calling a function:

```python
def remainder(number, divisor):
    return number % divisor

assert remainder(20, 7) == 6
```

All normal arguments to Python functions can also be passed by keyword, where the name of the argument is used in an assignment within the parentheses of a function call. The keyword arguments can be passed in any order as long as all of the required positional arguments are specified. You can mix and match keyword and positional arguments. These calls are equivalent:

```python
remainder(20, 7)
remainder(20, divisor=7)
remainder(number=20, divisor=7)
remainder(divisor=7, number=20)
```

Positional arguments must be specified before keyword arguments:

```python
remainder(number=20, 7)

>>>
Traceback ...
SyntaxError: positional argument follows keyword argument
```

Each argument can be specified only once:

```
remainder(20, number=7)
```

```
>>>
Traceback ...
TypeError: remainder() got multiple values for argument
➥'number'
```

If you already have a dictionary, and you want to use its contents to call a function like remainder, you can do this by using the ** operator. This instructs Python to pass the values from the dictionary as the corresponding keyword arguments of the function:

```
my_kwargs = {
    'number': 20,
    'divisor': 7,
}
assert remainder(**my_kwargs) == 6
```

You can mix the ** operator with positional arguments or keyword arguments in the function call, as long as no argument is repeated:

```
my_kwargs = {
    'divisor': 7,
}
assert remainder(number=20, **my_kwargs) == 6
```

You can also use the ** operator multiple times if you know that the dictionaries don't contain overlapping keys:

```
my_kwargs = {
    'number': 20,
}
other_kwargs = {
    'divisor': 7,
}
assert remainder(**my_kwargs, **other_kwargs) == 6
```

And if you'd like for a function to receive any named keyword argument, you can use the **kwargs catch-all parameter to collect those arguments into a dict that you can then process (see Item 26: "Define Function Decorators with functools.wraps" for when this is especially useful):

```
def print_parameters(**kwargs):
    for key, value in kwargs.items():
        print(f'{key} = {value}')

print_parameters(alpha=1.5, beta=9, gamma=4)
```

```
>>>
alpha = 1.5
beta = 9
gamma = 4
```

The flexibility of keyword arguments provides three significant benefits.

The first benefit is that keyword arguments make the function call clearer to new readers of the code. With the call remainder(20, 7), it's not evident which argument is number and which is divisor unless you look at the implementation of the remainder method. In the call with keyword arguments, number=20 and divisor=7 make it immediately obvious which parameter is being used for each purpose.

The second benefit of keyword arguments is that they can have default values specified in the function definition. This allows a function to provide additional capabilities when you need them, but you can accept the default behavior most of the time. This eliminates repetitive code and reduces noise.

For example, say that I want to compute the rate of fluid flowing into a vat. If the vat is also on a scale, then I could use the difference between two weight measurements at two different times to determine the flow rate:

```
def flow_rate(weight_diff, time_diff):
    return weight_diff / time_diff

weight_diff = 0.5
time_diff = 3
flow = flow_rate(weight_diff, time_diff)
print(f'{flow:.3} kg per second')
```

```
>>>
0.167 kg per second
```

In the typical case, it's useful to know the flow rate in kilograms per second. Other times, it'd be helpful to use the last sensor measurements to approximate larger time scales, like hours or days. I can provide this behavior in the same function by adding an argument for the time period scaling factor:

```
def flow_rate(weight_diff, time_diff, period):
    return (weight_diff / time_diff) * period
```

The problem is that now I need to specify the period argument every time I call the function, even in the common case of flow rate per second (where the period is 1):

```
flow_per_second = flow_rate(weight_diff, time_diff, 1)
```

To make this less noisy, I can give the period argument a default value:

```
def flow_rate(weight_diff, time_diff, period=1):
    return (weight_diff / time_diff) * period
```

The period argument is now optional:

```
flow_per_second = flow_rate(weight_diff, time_diff)
flow_per_hour = flow_rate(weight_diff, time_diff, period=3600)
```

This works well for simple default values; it gets tricky for complex default values (see Item 24: "Use None and Docstrings to Specify Dynamic Default Arguments" for details).

The third reason to use keyword arguments is that they provide a powerful way to extend a function's parameters while remaining backward compatible with existing callers. This means you can provide additional functionality without having to migrate a lot of existing code, which reduces the chance of introducing bugs.

For example, say that I want to extend the flow_rate function above to calculate flow rates in weight units besides kilograms. I can do this by adding a new optional parameter that provides a conversion rate to alternative measurement units:

```
def flow_rate(weight_diff, time_diff,
              period=1, units_per_kg=1):
    return ((weight_diff * units_per_kg) / time_diff) * period
```

The default argument value for units_per_kg is 1, which makes the returned weight units remain kilograms. This means that all existing callers will see no change in behavior. New callers to flow_rate can specify the new keyword argument to see the new behavior:

```
pounds_per_hour = flow_rate(weight_diff, time_diff,
                            period=3600, units_per_kg=2.2)
```

Providing backward compatibility using optional keyword arguments like this is also crucial for functions that accept *args (see Item 22: "Reduce Visual Noise with Variable Positional Arguments").

The only problem with this approach is that optional keyword arguments like period and units_per_kg may still be specified as positional arguments:

```
pounds_per_hour = flow_rate(weight_diff, time_diff, 3600, 2.2)
```

Supplying optional arguments positionally can be confusing because it isn't clear what the values 3600 and 2.2 correspond to. The best practice is to always specify optional arguments using the keyword names and never pass them as positional arguments. As a function author, you can also require that all callers use this more explicit keyword style to minimize potential errors (see Item 25: "Enforce Clarity with Keyword-Only and Positional-Only Arguments").

Things to Remember

✦ Function arguments can be specified by position or by keyword.

✦ Keywords make it clear what the purpose of each argument is when it would be confusing with only positional arguments.

✦ Keyword arguments with default values make it easy to add new behaviors to a function without needing to migrate all existing callers.

✦ Optional keyword arguments should always be passed by keyword instead of by position.

Item 24: Use None and Docstrings to Specify Dynamic Default Arguments

Sometimes you need to use a non-static type as a keyword argument's default value. For example, say I want to print logging messages that are marked with the time of the logged event. In the default case, I want the message to include the time when the function was called. I might try the following approach, assuming that the default arguments are reevaluated each time the function is called:

```
from time import sleep
from datetime import datetime

def log(message, when=datetime.now()):
    print(f'{when}: {message}')

log('Hi there!')
sleep(0.1)
log('Hello again!')

>>>
2019-07-06 14:06:15.120124: Hi there!
2019-07-06 14:06:15.120124: Hello again!
```

This doesn't work as expected. The timestamps are the same because datetime.now is executed only a single time: when the function is defined. A default argument value is evaluated only once per module load, which usually happens when a program starts up. After the module containing this code is loaded, the datetime.now() default argument will never be evaluated again.

The convention for achieving the desired result in Python is to provide a default value of None and to document the actual behavior in the docstring (see Item 84: "Write Docstrings for Every Function, Class, and Module" for background). When your code sees the argument value None, you allocate the default value accordingly:

```python
def log(message, when=None):
    """Log a message with a timestamp.

    Args:
        message: Message to print.
        when: datetime of when the message occurred.
            Defaults to the present time.
    """
    if when is None:
        when = datetime.now()
    print(f'{when}: {message}')
```

Now the timestamps will be different:

```python
log('Hi there!')
sleep(0.1)
log('Hello again!')
```

```
>>>
2019-07-06 14:06:15.222419: Hi there!
2019-07-06 14:06:15.322555: Hello again!
```

Using None for default argument values is especially important when the arguments are mutable. For example, say that I want to load a value encoded as JSON data; if decoding the data fails, I want an empty dictionary to be returned by default:

```python
import json


def decode(data, default={}):
    try:
        return json.loads(data)
    except ValueError:
        return default
```

The problem here is the same as in the datetime.now example above. The dictionary specified for default will be shared by all calls to decode because default argument values are evaluated only once (at module load time). This can cause extremely surprising behavior:

```
foo = decode('bad data')
foo['stuff'] = 5
bar = decode('also bad')
bar['meep'] = 1
print('Foo:', foo)
print('Bar:', bar)

>>>
Foo: {'stuff': 5, 'meep': 1}
Bar: {'stuff': 5, 'meep': 1}
```

You might expect two different dictionaries, each with a single key and value. But modifying one seems to also modify the other. The culprit is that foo and bar are both equal to the default parameter. They are the same dictionary object:

```
assert foo is bar
```

The fix is to set the keyword argument default value to None and then document the behavior in the function's docstring:

```
def decode(data, default=None):
    """Load JSON data from a string.

    Args:
        data: JSON data to decode.
        default: Value to return if decoding fails.
            Defaults to an empty dictionary.
    """
    try:
        return json.loads(data)
    except ValueError:
        if default is None:
            default = {}
        return default
```

Now, running the same test code as before produces the expected result:

```
foo = decode('bad data')
foo['stuff'] = 5
bar = decode('also bad')
bar['meep'] = 1
print('Foo:', foo)
print('Bar:', bar)
assert foo is not bar
```

```
>>>
Foo: {'stuff': 5}
Bar: {'meep': 1}
```

This approach also works with type annotations (see Item 90: "Consider Static Analysis via typing to Obviate Bugs"). Here, the when argument is marked as having an Optional value that is a datetime. Thus, the only two valid choices for when are None or a datetime object:

```
from typing import Optional

def log_typed(message: str,
              when: Optional[datetime]=None) -> None:
    """Log a message with a timestamp.

    Args:
        message: Message to print.
        when: datetime of when the message occurred.
            Defaults to the present time.
    """
    if when is None:
        when = datetime.now()
    print(f'{when}: {message}')
```

Things to Remember

✦ A default argument value is evaluated only once: during function definition at module load time. This can cause odd behaviors for dynamic values (like {}, [], or datetime.now()).

✦ Use None as the default value for any keyword argument that has a dynamic value. Document the actual default behavior in the function's docstring.

✦ Using None to represent keyword argument default values also works correctly with type annotations.

Item 25: Enforce Clarity with Keyword-Only and Positional-Only Arguments

Passing arguments by keyword is a powerful feature of Python functions (see Item 23: "Provide Optional Behavior with Keyword Arguments"). The flexibility of keyword arguments enables you to write functions that will be clear to new readers of your code for many use cases.

For example, say I want to divide one number by another but know that I need to be very careful about special cases. Sometimes, I want to ignore ZeroDivisionError exceptions and return infinity instead. Other times, I want to ignore OverflowError exceptions and return zero instead:

```
def safe_division(number, divisor,
                  ignore_overflow,
                  ignore_zero_division):
    try:
        return number / divisor
    except OverflowError:
        if ignore_overflow:
            return 0
        else:
            raise
    except ZeroDivisionError:
        if ignore_zero_division:
            return float('inf')
        else:
            raise
```

Using this function is straightforward. This call ignores the float overflow from division and returns zero:

```
result = safe_division(1.0, 10**500, True, False)
print(result)

>>>
0
```

This call ignores the error from dividing by zero and returns infinity:

```
result = safe_division(1.0, 0, False, True)
print(result)

>>>
inf
```

The problem is that it's easy to confuse the position of the two Boolean arguments that control the exception-ignoring behavior. This can easily cause bugs that are hard to track down. One way to improve the readability of this code is to use keyword arguments. By default, the function can be overly cautious and can always re-raise exceptions:

```
def safe_division_b(number, divisor,
                    ignore_overflow=False,       # Changed
                    ignore_zero_division=False): # Changed
    ...
```

Then, callers can use keyword arguments to specify which of the ignore flags they want to set for specific operations, overriding the default behavior:

```
result = safe_division_b(1.0, 10**500, ignore_overflow=True)
print(result)

result = safe_division_b(1.0, 0, ignore_zero_division=True)
print(result)

>>>
0
inf
```

The problem is, since these keyword arguments are optional behavior, there's nothing forcing callers to use keyword arguments for clarity. Even with the new definition of safe_division_b, you can still call it the old way with positional arguments:

```
assert safe_division_b(1.0, 10**500, True, False) == 0
```

With complex functions like this, it's better to require that callers are clear about their intentions by defining functions with *keyword-only arguments*. These arguments can only be supplied by keyword, never by position.

Here, I redefine the safe_division function to accept keyword-only arguments. The * symbol in the argument list indicates the end of positional arguments and the beginning of keyword-only arguments:

```
def safe_division_c(number, divisor, *,   # Changed
                    ignore_overflow=False,
                    ignore_zero_division=False):
    ...
```

Now, calling the function with positional arguments for the keyword arguments won't work:

```
safe_division_c(1.0, 10**500, True, False)

>>>
Traceback ...
TypeError: safe_division_c() takes 2 positional arguments but 4
➥were given
```

But keyword arguments and their default values will work as expected (ignoring an exception in one case and raising it in another):

```
result = safe_division_c(1.0, 0, ignore_zero_division=True)
assert result == float('inf')
```

```
try:
    result = safe_division_c(1.0, 0)
except ZeroDivisionError:
    pass  # Expected
```

However, a problem still remains with the safe_division_c version of this function: Callers may specify the first two required arguments (number and divisor) with a mix of positions and keywords:

```
assert safe_division_c(number=2, divisor=5) == 0.4
assert safe_division_c(divisor=5, number=2) == 0.4
assert safe_division_c(2, divisor=5) == 0.4
```

Later, I may decide to change the names of these first two arguments because of expanding needs or even just because my style preferences change:

```
def safe_division_c(numerator, denominator, *,  # Changed
                    ignore_overflow=False,
                    ignore_zero_division=False):
    ...
```

Unfortunately, this seemingly superficial change breaks all the existing callers that specified the number or divisor arguments using keywords:

```
safe_division_c(number=2, divisor=5)
```

```
>>>
Traceback ...
TypeError: safe_division_c() got an unexpected keyword argument
➥'number'
```

This is especially problematic because I never intended for number and divisor to be part of an explicit interface for this function. These were just convenient parameter names that I chose for the implementation, and I didn't expect anyone to rely on them explicitly.

Python 3.8 introduces a solution to this problem, called *positional-only arguments*. These arguments can be supplied only by position and never by keyword (the opposite of the keyword-only arguments demonstrated above).

Here, I redefine the safe_division function to use positional-only arguments for the first two required parameters. The / symbol in the argument list indicates where positional-only arguments end:

```
def safe_division_d(numerator, denominator, /, *,  # Changed
                    ignore_overflow=False,
                    ignore_zero_division=False):
    ...
```

I can verify that this function works when the required arguments are provided positionally:

```
assert safe_division_d(2, 5) == 0.4
```

But an exception is raised if keywords are used for the positional-only parameters:

```
safe_division_d(numerator=2, denominator=5)
```

```
>>>
Traceback ...
TypeError: safe_division_d() got some positional-only arguments
➥passed as keyword arguments: 'numerator, denominator'
```

Now, I can be sure that the first two required positional arguments in the definition of the safe_division_d function are decoupled from callers. I won't break anyone if I change the parameters' names again.

One notable consequence of keyword- and positional-only arguments is that any parameter name between the / and * symbols in the argument list may be passed either by position or by keyword (which is the default for all function arguments in Python). Depending on your API's style and needs, allowing both argument passing styles can increase readability and reduce noise. For example, here I've added another optional parameter to safe_division that allows callers to specify how many digits to use in rounding the result:

```
def safe_division_e(numerator, denominator, /,
                    ndigits=10, *,                    # Changed
                    ignore_overflow=False,
                    ignore_zero_division=False):
    try:
        fraction = numerator / denominator            # Changed
        return round(fraction, ndigits)               # Changed
    except OverflowError:
        if ignore_overflow:
            return 0
        else:
            raise
    except ZeroDivisionError:
        if ignore_zero_division:
            return float('inf')
        else:
            raise
```

Now, I can call this new version of the function in all these different ways, since ndigits is an optional parameter that may be passed either by position or by keyword:

```
result = safe_division_e(22, 7)
print(result)

result = safe_division_e(22, 7, 5)
print(result)

result = safe_division_e(22, 7, ndigits=2)
print(result)

>>>
3.1428571429
3.14286
3.14
```

Things to Remember

+ Keyword-only arguments force callers to supply certain arguments by keyword (instead of by position), which makes the intention of a function call clearer. Keyword-only arguments are defined after a single * in the argument list.

+ Positional-only arguments ensure that callers can't supply certain parameters using keywords, which helps reduce coupling. Positional-only arguments are defined before a single / in the argument list.

+ Parameters between the / and * characters in the argument list may be supplied by position or keyword, which is the default for Python parameters.

Item 26: Define Function Decorators with functools.wraps

Python has special syntax for *decorators* that can be applied to functions. A decorator has the ability to run additional code before and after each call to a function it wraps. This means decorators can access and modify input arguments, return values, and raised exceptions. This functionality can be useful for enforcing semantics, debugging, registering functions, and more.

For example, say that I want to print the arguments and return value of a function call. This can be especially helpful when debugging the stack of nested function calls from a recursive function. Here, I define such a decorator by using *args and **kwargs (see Item 22: "Reduce Visual Noise with Variable Positional Arguments" and Item 23: "Provide Optional Behavior with Keyword Arguments") to pass through all parameters to the wrapped function:

```
def trace(func):
    def wrapper(*args, **kwargs):
        result = func(*args, **kwargs)
        print(f'{func.__name__}({args!r}, {kwargs!r}) '
              f'-> {result!r}')
        return result
    return wrapper
```

I can apply this decorator to a function by using the @ symbol:

```
@trace
def fibonacci(n):
    """Return the n-th Fibonacci number"""
    if n in (0, 1):
        return n
    return (fibonacci(n - 2) + fibonacci(n - 1))
```

Using the @ symbol is equivalent to calling the decorator on the function it wraps and assigning the return value to the original name in the same scope:

```
fibonacci = trace(fibonacci)
```

The decorated function runs the wrapper code before and after fibonacci runs. It prints the arguments and return value at each level in the recursive stack:

```
fibonacci(4)
```

```
>>>
fibonacci((0,), {}) -> 0
fibonacci((1,), {}) -> 1
fibonacci((2,), {}) -> 1
fibonacci((1,), {}) -> 1
fibonacci((0,), {}) -> 0
fibonacci((1,), {}) -> 1
fibonacci((2,), {}) -> 1
fibonacci((3,), {}) -> 2
fibonacci((4,), {}) -> 3
```

This works well, but it has an unintended side effect. The value returned by the decorator—the function that's called above—doesn't think it's named fibonacci:

```
print(fibonacci)
```

```
>>>
<function trace.<locals>.wrapper at 0x108955dc0>
```

The cause of this isn't hard to see. The trace function returns the wrapper defined within its body. The wrapper function is what's assigned to the fibonacci name in the containing module because of the decorator. This behavior is problematic because it undermines tools that do introspection, such as debuggers (see Item 80: "Consider Interactive Debugging with pdb").

For example, the help built-in function is useless when called on the decorated fibonacci function. It should instead print out the docstring defined above ('Return the n-th Fibonacci number'):

```
help(fibonacci)
```

```
>>>
Help on function wrapper in module __main__:
```

```
wrapper(*args, **kwargs)
```

Object serializers (see Item 68: "Make pickle Reliable with copyreg") break because they can't determine the location of the original function that was decorated:

```
import pickle
```

```
pickle.dumps(fibonacci)
```

```
>>>
Traceback ...
AttributeError: Can't pickle local object 'trace.<locals>.
➡wrapper'
```

The solution is to use the wraps helper function from the functools built-in module. This is a decorator that helps you write decorators. When you apply it to the wrapper function, it copies all of the important metadata about the inner function to the outer function:

```
from functools import wraps
```

```
def trace(func):
    @wraps(func)
    def wrapper(*args, **kwargs):
        ...
    return wrapper

@trace
def fibonacci(n):
    ...
```

Now, running the `help` function produces the expected result, even though the function is decorated:

```
help(fibonacci)
```

```
>>>
Help on function fibonacci in module __main__:

fibonacci(n)
    Return the n-th Fibonacci number
```

The `pickle` object serializer also works:

```
print(pickle.dumps(fibonacci))
```

```
>>>
b'\x80\x04\x95\x1a\x00\x00\x00\x00\x00\x00\x00\x8c\x08__main__\
➥x94\x8c\tfibonacci\x94\x93\x94.'
```

Beyond these examples, Python functions have many other standard attributes (e.g., `__name__`, `__module__`, `__annotations__`) that must be preserved to maintain the interface of functions in the language. Using `wraps` ensures that you'll always get the correct behavior.

Things to Remember

✦ Decorators in Python are syntax to allow one function to modify another function at runtime.

✦ Using decorators can cause strange behaviors in tools that do introspection, such as debuggers.

✦ Use the `wraps` decorator from the `functools` built-in module when you define your own decorators to avoid issues.

Chapter 4

Comprehensions and Generators

Many programs are built around processing lists, dictionary key/value pairs, and sets. Python provides a special syntax, called *comprehensions*, for succinctly iterating through these types and creating derivative data structures. Comprehensions can significantly increase the readability of code performing these common tasks and provide a number of other benefits.

This style of processing is extended to functions with *generators*, which enable a stream of values to be incrementally returned by a function. The result of a call to a generator function can be used anywhere an iterator is appropriate (e.g., for loops, starred expressions). Generators can improve performance, reduce memory usage, and increase readability.

Item 27: Use Comprehensions Instead of map and filter

Python provides compact syntax for deriving a new list from another sequence or iterable. These expressions are called *list comprehensions*. For example, say that I want to compute the square of each number in a list. Here, I do this by using a simple for loop:

```
a = [1, 2, 3, 4, 5, 6, 7, 8, 9, 10]
squares = []
for x in a:
    squares.append(x**2)
print(squares)

>>>
[1, 4, 9, 16, 25, 36, 49, 64, 81, 100]
```

With a list comprehension, I can achieve the same outcome by specifying the expression for my computation along with the input sequence to loop over:

```
squares = [x**2 for x in a]   # List comprehension
print(squares)
```

```
>>>
[1, 4, 9, 16, 25, 36, 49, 64, 81, 100]
```

Unless you're applying a single-argument function, list comprehensions are also clearer than the map built-in function for simple cases. map requires the creation of a lambda function for the computation, which is visually noisy:

```
alt = map(lambda x: x ** 2, a)
```

Unlike map, list comprehensions let you easily filter items from the input list, removing corresponding outputs from the result. For example, say I want to compute the squares of the numbers that are divisible by 2. Here, I do this by adding a conditional expression to the list comprehension after the loop:

```
even_squares = [x**2 for x in a if x % 2 == 0]
print(even_squares)
```

```
>>>
[4, 16, 36, 64, 100]
```

The filter built-in function can be used along with map to achieve the same outcome, but it is much harder to read:

```
alt = map(lambda x: x**2, filter(lambda x: x % 2 == 0, a))
assert even_squares == list(alt)
```

Dictionaries and sets have their own equivalents of list comprehensions (called *dictionary comprehensions* and *set comprehensions*, respectively). These make it easy to create other types of derivative data structures when writing algorithms:

```
even_squares_dict = {x: x**2 for x in a if x % 2 == 0}
threes_cubed_set = {x**3 for x in a if x % 3 == 0}
print(even_squares_dict)
print(threes_cubed_set)
```

```
>>>
{2: 4, 4: 16, 6: 36, 8: 64, 10: 100}
{216, 729, 27}
```

Achieving the same outcome is possible with map and filter if you wrap each call with a corresponding constructor. These statements

get so long that you have to break them up across multiple lines,
which is even noisier and should be avoided:

```
alt_dict = dict(map(lambda x: (x, x**2),
             filter(lambda x: x % 2 == 0, a)))
alt_set = set(map(lambda x: x**3,
             filter(lambda x: x % 3 == 0, a)))
```

Things to Remember

✦ List comprehensions are clearer than the map and filter built-in
functions because they don't require lambda expressions.

✦ List comprehensions allow you to easily skip items from the input
list, a behavior that map doesn't support without help from filter.

✦ Dictionaries and sets may also be created using comprehensions.

Item 28: Avoid More Than Two Control Subexpressions in Comprehensions

Beyond basic usage (see Item 27: "Use Comprehensions Instead of map
and filter"), comprehensions support multiple levels of looping. For
example, say that I want to simplify a matrix (a list containing other
list instances) into one flat list of all cells. Here, I do this with a list
comprehension by including two for subexpressions. These subexpressions run in the order provided, from left to right:

```
matrix = [[1, 2, 3], [4, 5, 6], [7, 8, 9]]
flat = [x for row in matrix for x in row]
print(flat)
```

```
>>>
[1, 2, 3, 4, 5, 6, 7, 8, 9]
```

This example is simple, readable, and a reasonable usage of multiple
loops in a comprehension. Another reasonable usage of multiple loops
involves replicating the two-level-deep layout of the input list. For
example, say that I want to square the value in each cell of a two-
dimensional matrix. This comprehension is noisier because of the
extra [] characters, but it's still relatively easy to read:

```
squared = [[x**2 for x in row] for row in matrix]
print(squared)
```

```
>>>
[[1, 4, 9], [16, 25, 36], [49, 64, 81]]
```

If this comprehension included another loop, it would get so long that I'd have to split it over multiple lines:

```
my_lists = [
    [[1, 2, 3], [4, 5, 6]],
    ...
]
flat = [x for sublist1 in my_lists
          for sublist2 in sublist1
          for x in sublist2]
```

At this point, the multiline comprehension isn't much shorter than the alternative. Here, I produce the same result using normal loop statements. The indentation of this version makes the looping clearer than the three-level-list comprehension:

```
flat = []
for sublist1 in my_lists:
    for sublist2 in sublist1:
        flat.extend(sublist2)
```

Comprehensions support multiple if conditions. Multiple conditions at the same loop level have an implicit and expression. For example, say that I want to filter a list of numbers to only even values greater than 4. These two list comprehensions are equivalent:

```
a = [1, 2, 3, 4, 5, 6, 7, 8, 9, 10]
b = [x for x in a if x > 4 if x % 2 == 0]
c = [x for x in a if x > 4 and x % 2 == 0]
```

Conditions can be specified at each level of looping after the for subexpression. For example, say I want to filter a matrix so the only cells remaining are those divisible by 3 in rows that sum to 10 or higher. Expressing this with a list comprehension does not require a lot of code, but it is extremely difficult to read:

```
matrix = [[1, 2, 3], [4, 5, 6], [7, 8, 9]]
filtered = [[x for x in row if x % 3 == 0]
              for row in matrix if sum(row) >= 10]
print(filtered)

>>>
[[6], [9]]
```

Although this example is a bit convoluted, in practice you'll see situations arise where such comprehensions seem like a good fit. I strongly encourage you to avoid using list, dict, or set comprehensions that look like this. The resulting code is very difficult for new readers to understand. The potential for confusion is even worse for dict comprehensions since they already need an extra parameter to represent both the key and the value for each item.

The rule of thumb is to avoid using more than two control subexpressions in a comprehension. This could be two conditions, two loops, or one condition and one loop. As soon as it gets more complicated than that, you should use normal if and for statements and write a helper function (see Item 30: "Consider Generators Instead of Returning Lists").

Things to Remember

✦ Comprehensions support multiple levels of loops and multiple conditions per loop level.

✦ Comprehensions with more than two control subexpressions are very difficult to read and should be avoided.

Item 29: Avoid Repeated Work in Comprehensions by Using Assignment Expressions

A common pattern with comprehensions—including list, dict, and set variants—is the need to reference the same computation in multiple places. For example, say that I'm writing a program to manage orders for a fastener company. As new orders come in from customers, I need to be able to tell them whether I can fulfill their orders. I need to verify that a request is sufficiently in stock and above the minimum threshold for shipping (in batches of 8):

```
stock = {
    'nails': 125,
    'screws': 35,
    'wingnuts': 8,
    'washers': 24,
}

order = ['screws', 'wingnuts', 'clips']

def get_batches(count, size):
    return count // size

result = {}
for name in order:
  count = stock.get(name, 0)
  batches = get_batches(count, 8)
```

```
 if batches:
    result[name] = batches

print(result)

>>>
{'screws': 4, 'wingnuts': 1}
```

Here, I implement this looping logic more succinctly using a dictionary comprehension (see Item 27: "Use Comprehensions Instead of map and filter" for best practices):

```
found = {name: get_batches(stock.get(name, 0), 8)
         for name in order
         if get_batches(stock.get(name, 0), 8)}
print(found)

>>>
{'screws': 4, 'wingnuts': 1}
```

Although this code is more compact, the problem with it is that the get_batches(stock.get(name, 0), 8) expression is repeated. This hurts readability by adding visual noise that's technically unnecessary. It also increases the likelihood of introducing a bug if the two expressions aren't kept in sync. For example, here I've changed the first get_batches call to have 4 as its second parameter instead of 8, which causes the results to be different:

```
has_bug = {name: get_batches(stock.get(name, 0), 4)
           for name in order
           if get_batches(stock.get(name, 0), 8)}

print('Expected:', found)
print('Found:   ', has_bug)

>>>
Expected: {'screws': 4, 'wingnuts': 1}
Found:    {'screws': 8, 'wingnuts': 2}
```

An easy solution to these problems is to use the walrus operator (:=), which was introduced in Python 3.8, to form an assignment expression as part of the comprehension (see Item 10: "Prevent Repetition with Assignment Expressions" for background):

```
found = {name: batches for name in order
         if (batches := get_batches(stock.get(name, 0), 8))}
```

The assignment expression (batches := get_batches(...)) allows me to look up the value for each order key in the stock dictionary a single time, call get_batches once, and then store its corresponding value in the batches variable. I can then reference that variable elsewhere in the comprehension to construct the dict's contents instead of having to call get_batches a second time. Eliminating the redundant calls to get and get_batches may also improve performance by avoiding unnecessary computations for each item in the order list.

It's valid syntax to define an assignment expression in the value expression for a comprehension. But if you try to reference the variable it defines in other parts of the comprehension, you might get an exception at runtime because of the order in which comprehensions are evaluated:

```
result = {name: (tenth := count // 10)
          for name, count in stock.items() if tenth > 0}
```

```
>>>
Traceback ...
NameError: name 'tenth' is not defined
```

I can fix this example by moving the assignment expression into the condition and then referencing the variable name it defined in the comprehension's value expression:

```
result = {name: tenth for name, count in stock.items()
          if (tenth := count // 10) > 0}
print(result)
```

```
>>>
{'nails': 12, 'screws': 3, 'washers': 2}
```

If a comprehension uses the walrus operator in the value part of the comprehension and doesn't have a condition, it'll leak the loop variable into the containing scope (see Item 21: "Know How Closures Interact with Variable Scope" for background):

```
half = [(last := count // 2) for count in stock.values()]
print(f'Last item of {half} is {last}')
```

```
>>>
Last item of [62, 17, 4, 12] is 12
```

This leakage of the loop variable is similar to what happens with a normal for loop:

```
for count in stock.values():  # Leaks loop variable
    pass
print(f'Last item of {list(stock.values())} is {count}')
```

```
>>>
Last item of [125, 35, 8, 24] is 24
```

However, similar leakage doesn't happen for the loop variables from comprehensions:

```
half = [count // 2 for count in stock.values()]
print(half)   # Works
print(count)  # Exception because loop variable didn't leak
```

```
>>>
[62, 17, 4, 12]
Traceback ...
NameError: name 'count' is not defined
```

It's better not to leak loop variables, so I recommend using assignment expressions only in the condition part of a comprehension.

Using an assignment expression also works the same way in generator expressions (see Item 32: "Consider Generator Expressions for Large List Comprehensions"). Here, I create an iterator of pairs containing the item name and the current count in stock instead of a dict instance:

```
found = ((name, batches) for name in order
         if (batches := get_batches(stock.get(name, 0), 8)))
print(next(found))
print(next(found))
```

```
>>>
('screws', 4)
('wingnuts', 1)
```

Things to Remember

✦ Assignment expressions make it possible for comprehensions and generator expressions to reuse the value from one condition elsewhere in the same comprehension, which can improve readability and performance.

✦ Although it's possible to use an assignment expression outside of a comprehension or generator expression's condition, you should avoid doing so.

Item 30: Consider Generators Instead of Returning Lists

The simplest choice for a function that produces a sequence of results is to return a list of items. For example, say that I want to find the index of every word in a string. Here, I accumulate results in a list using the append method and return it at the end of the function:

```
def index_words(text):
    result = []
    if text:
        result.append(0)
    for index, letter in enumerate(text):
        if letter == ' ':
            result.append(index + 1)
    return result
```

This works as expected for some sample input:

```
address = 'Four score and seven years ago...'
result = index_words(address)
print(result[:10])
```

```
>>>
[0, 5, 11, 15, 21, 27, 31, 35, 43, 51]
```

There are two problems with the index_words function.

The first problem is that the code is a bit dense and noisy. Each time a new result is found, I call the append method. The method call's bulk (result.append) deemphasizes the value being added to the list (index + 1). There is one line for creating the result list and another for returning it. While the function body contains ~130 characters (without whitespace), only ~75 characters are important.

A better way to write this function is by using a *generator*. Generators are produced by functions that use yield expressions. Here, I define a generator function that produces the same results as before:

```
def index_words_iter(text):
    if text:
        yield 0
    for index, letter in enumerate(text):
        if letter == ' ':
            yield index + 1
```

When called, a generator function does not actually run but instead immediately returns an iterator. With each call to the next built-in function, the iterator advances the generator to its next yield expression. Each value passed to yield by the generator is returned by the iterator to the caller:

```
it = index_words_iter(address)
print(next(it))
print(next(it))
```

```
>>>
0
5
```

The `index_words_iter` function is significantly easier to read because all interactions with the result list have been eliminated. Results are passed to `yield` expressions instead. You can easily convert the iterator returned by the generator to a list by passing it to the `list` built-in function if necessary (see Item 32: "Consider Generator Expressions for Large List Comprehensions" for how this works):

```
result = list(index_words_iter(address))
print(result[:10])
```

```
>>>
[0, 5, 11, 15, 21, 27, 31, 35, 43, 51]
```

The second problem with `index_words` is that it requires all results to be stored in the list before being returned. For huge inputs, this can cause a program to run out of memory and crash.

In contrast, a generator version of this function can easily be adapted to take inputs of arbitrary length due to its bounded memory requirements. For example, here I define a generator that streams input from a file one line at a time and yields outputs one word at a time:

```
def index_file(handle):
    offset = 0
    for line in handle:
        if line:
            yield offset
        for letter in line:
            offset += 1
            if letter == ' ':
                yield offset
```

The working memory for this function is limited to the maximum length of one line of input. Running the generator produces the same results (see Item 36: "Consider `itertools` for Working with Iterators and Generators" for more about the `islice` function):

```
with open('address.txt', 'r') as f:
    it = index_file(f)
    results = itertools.islice(it, 0, 10)
    print(list(results))
```

```
>>>
[0, 5, 11, 15, 21, 27, 31, 35, 43, 51]
```

The only gotcha with defining generators like this is that the callers must be aware that the iterators returned are stateful and can't be reused (see Item 31: "Be Defensive When Iterating Over Arguments").

Things to Remember

✦ Using generators can be clearer than the alternative of having a function return a `list` of accumulated results.

✦ The iterator returned by a generator produces the set of values passed to `yield` expressions within the generator function's body.

✦ Generators can produce a sequence of outputs for arbitrarily large inputs because their working memory doesn't include all inputs and outputs.

Item 31: Be Defensive When Iterating Over Arguments

When a function takes a `list` of objects as a parameter, it's often important to iterate over that `list` multiple times. For example, say that I want to analyze tourism numbers for the U.S. state of Texas. Imagine that the data set is the number of visitors to each city (in millions per year). I'd like to figure out what percentage of overall tourism each city receives.

To do this, I need a normalization function that sums the inputs to determine the total number of tourists per year and then divides each city's individual visitor count by the total to find that city's contribution to the whole:

```
def normalize(numbers):
    total = sum(numbers)
    result = []
    for value in numbers:
        percent = 100 * value / total
        result.append(percent)
    return result
```

This function works as expected when given a list of visits:

```
visits = [15, 35, 80]
percentages = normalize(visits)
print(percentages)
assert sum(percentages) == 100.0

>>>
[11.538461538461538, 26.923076923076923, 61.53846153846154]
```

To scale this up, I need to read the data from a file that contains every city in all of Texas. I define a generator to do this because then I can reuse the same function later, when I want to compute tourism numbers for the whole world—a much larger data set with higher memory requirements (see Item 30: "Consider Generators Instead of Returning Lists" for background):

```
def read_visits(data_path):
    with open(data_path) as f:
        for line in f:
            yield int(line)
```

Surprisingly, calling `normalize` on the `read_visits` generator's return value produces no results:

```
it = read_visits('my_numbers.txt')
percentages = normalize(it)
print(percentages)
```

```
>>>
[]
```

This behavior occurs because an iterator produces its results only a single time. If you iterate over an iterator or a generator that has already raised a `StopIteration` exception, you won't get any results the second time around:

```
it = read_visits('my_numbers.txt')
print(list(it))
print(list(it))  # Already exhausted
```

```
>>>
[15, 35, 80]
[]
```

Confusingly, you also won't get errors when you iterate over an already exhausted iterator. `for` loops, the `list` constructor, and many other functions throughout the Python standard library expect the `StopIteration` exception to be raised during normal operation. These functions can't tell the difference between an iterator that has no output and an iterator that had output and is now exhausted.

To solve this problem, you can explicitly exhaust an input iterator and keep a copy of its entire contents in a `list`. You can then iterate over the `list` version of the data as many times as you need to. Here's the same function as before, but it defensively copies the input iterator:

```
def normalize_copy(numbers):
    numbers_copy = list(numbers)  # Copy the iterator
```

```
    total = sum(numbers_copy)
    result = []
    for value in numbers_copy:
        percent = 100 * value / total
        result.append(percent)
    return result
```

Now the function works correctly on the read_visits generator's return value:

```
it = read_visits('my_numbers.txt')
percentages = normalize_copy(it)
print(percentages)
assert sum(percentages) == 100.0
```

```
>>>
[11.538461538461538, 26.923076923076923, 61.53846153846154]
```

The problem with this approach is that the copy of the input iterator's contents could be extremely large. Copying the iterator could cause the program to run out of memory and crash. This potential for scalability issues undermines the reason that I wrote read_visits as a generator in the first place. One way around this is to accept a function that returns a new iterator each time it's called:

```
def normalize_func(get_iter):
    total = sum(get_iter())    # New iterator
    result = []
    for value in get_iter():   # New iterator
        percent = 100 * value / total
        result.append(percent)
    return result
```

To use normalize_func, I can pass in a lambda expression that calls the generator and produces a new iterator each time:

```
path = 'my_numbers.txt'
percentages = normalize_func(lambda: read_visits(path))
print(percentages)
assert sum(percentages) == 100.0
```

```
>>>
[11.538461538461538, 26.923076923076923, 61.53846153846154]
```

Although this works, having to pass a lambda function like this is clumsy. A better way to achieve the same result is to provide a new container class that implements the *iterator protocol*.

The iterator protocol is how Python for loops and related expressions traverse the contents of a container type. When Python sees a statement like for x in foo, it actually calls iter(foo). The iter built-in function calls the foo.__iter__ special method in turn. The __iter__ method must return an iterator object (which itself implements the __next__ special method). Then, the for loop repeatedly calls the next built-in function on the iterator object until it's exhausted (indicated by raising a StopIteration exception).

It sounds complicated, but practically speaking, you can achieve all of this behavior for your classes by implementing the __iter__ method as a generator. Here, I define an iterable container class that reads the file containing tourism data:

```
class ReadVisits:
    def __init__(self, data_path):
        self.data_path = data_path

    def __iter__(self):
        with open(self.data_path) as f:
            for line in f:
                yield int(line)
```

This new container type works correctly when passed to the original function without modifications:

```
visits = ReadVisits(path)
percentages = normalize(visits)
print(percentages)
assert sum(percentages) == 100.0

>>>
[11.538461538461538, 26.923076923076923, 61.53846153846154]
```

This works because the sum method in normalize calls ReadVisits. __iter__ to allocate a new iterator object. The for loop to normalize the numbers also calls __iter__ to allocate a second iterator object. Each of those iterators will be advanced and exhausted independently, ensuring that each unique iteration sees all of the input data values. The only downside of this approach is that it reads the input data multiple times.

Now that you know how containers like ReadVisits work, you can write your functions and methods to ensure that parameters aren't just iterators. The protocol states that when an iterator is passed to the iter built-in function, iter returns the iterator itself. In contrast, when a container type is passed to iter, a new iterator object is returned each time. Thus, you can test an input value for this behavior and raise a TypeError to reject arguments that can't be repeatedly iterated over:

```
def normalize_defensive(numbers):
    if iter(numbers) is numbers:  # An iterator -- bad!
        raise TypeError('Must supply a container')
    total = sum(numbers)
    result = []
    for value in numbers:
        percent = 100 * value / total
        result.append(percent)
    return result
```

Alternatively, the collections.abc built-in module defines an Iterator class that can be used in an isinstance test to recognize the potential problem (see Item 43: "Inherit from collections.abc for Custom Container Types"):

```
from collections.abc import Iterator


def normalize_defensive(numbers):
    if isinstance(numbers, Iterator):  # Another way to check
        raise TypeError('Must supply a container')
    total = sum(numbers)
    result = []
    for value in numbers:
        percent = 100 * value / total
        result.append(percent)
    return result
```

The approach of using a container is ideal if you don't want to copy the full input iterator, as with the normalize_copy function above, but you also need to iterate over the input data multiple times. This function works as expected for list and ReadVisits inputs because they are iterable containers that follow the iterator protocol:

```
visits = [15, 35, 80]
percentages = normalize_defensive(visits)
assert sum(percentages) == 100.0

visits = ReadVisits(path)
percentages = normalize_defensive(visits)
assert sum(percentages) == 100.0
```

The function raises an exception if the input is an iterator rather than a container:

```
visits = [15, 35, 80]
it = iter(visits)
normalize_defensive(it)
```

```
>>>
Traceback ...
TypeError: Must supply a container
```

The same approach can also be used for asynchronous iterators (see Item 61: "Know How to Port Threaded I/O to asyncio" for an example).

Things to Remember

✦ Beware of functions and methods that iterate over input arguments multiple times. If these arguments are iterators, you may see strange behavior and missing values.

✦ Python's iterator protocol defines how containers and iterators interact with the iter and next built-in functions, for loops, and related expressions.

✦ You can easily define your own iterable container type by implementing the __iter__ method as a generator.

✦ You can detect that a value is an iterator (instead of a container) if calling iter on it produces the same value as what you passed in. Alternatively, you can use the isinstance built-in function along with the collections.abc.Iterator class.

Item 32: Consider Generator Expressions for Large List Comprehensions

The problem with list comprehensions (see Item 27: "Use Comprehensions Instead of map and filter") is that they may create new list instances containing one item for each value in input sequences. This is fine for small inputs, but for large inputs, this behavior could consume significant amounts of memory and cause a program to crash.

For example, say that I want to read a file and return the number of characters on each line. Doing this with a list comprehension would require holding the length of every line of the file in memory. If the file is enormous or perhaps a never-ending network socket, using list comprehensions would be problematic. Here, I use a list comprehension in a way that can only handle small input values:

```
value = [len(x) for x in open('my_file.txt')]
print(value)
```

```
>>>
[100, 57, 15, 1, 12, 75, 5, 86, 89, 11]
```

To solve this issue, Python provides *generator expressions*, which are a generalization of list comprehensions and generators. Generator expressions don't materialize the whole output sequence when they're run. Instead, generator expressions evaluate to an iterator that yields one item at a time from the expression.

You create a generator expression by putting list-comprehension-like syntax between () characters. Here, I use a generator expression that is equivalent to the code above. However, the generator expression immediately evaluates to an iterator and doesn't make forward progress:

```
it = (len(x) for x in open('my_file.txt'))
print(it)
```

```
>>>
<generator object <genexpr> at 0x108993dd0>
```

The returned iterator can be advanced one step at a time to produce the next output from the generator expression, as needed (using the next built-in function). I can consume as much of the generator expression as I want without risking a blowup in memory usage:

```
print(next(it))
print(next(it))
```

```
>>>
100
57
```

Another powerful outcome of generator expressions is that they can be composed together. Here, I take the iterator returned by the generator expression above and use it as the input for another generator expression:

```
roots = ((x, x**0.5) for x in it)
```

Each time I advance this iterator, it also advances the interior iterator, creating a domino effect of looping, evaluating conditional expressions, and passing around inputs and outputs, all while being as memory efficient as possible:

```
print(next(roots))
```

```
>>>
(15, 3.872983346207417)
```

Chaining generators together like this executes very quickly in Python. When you're looking for a way to compose functionality that's operating on a large stream of input, generator expressions are a great choice. The only gotcha is that the iterators returned by generator expressions are stateful, so you must be careful not to use these iterators more than once (see Item 31: "Be Defensive When Iterating Over Arguments").

Things to Remember

✦ List comprehensions can cause problems for large inputs by using too much memory.

✦ Generator expressions avoid memory issues by producing outputs one at a time as iterators.

✦ Generator expressions can be composed by passing the iterator from one generator expression into the for subexpression of another.

✦ Generator expressions execute very quickly when chained together and are memory efficient.

Item 33: Compose Multiple Generators with yield from

Generators provide a variety of benefits (see Item 30: "Consider Generators Instead of Returning Lists") and solutions to common problems (see Item 31: "Be Defensive When Iterating Over Arguments"). Generators are so useful that many programs start to look like layers of generators strung together.

For example, say that I have a graphical program that's using generators to animate the movement of images onscreen. To get the visual effect I'm looking for, I need the images to move quickly at first, pause temporarily, and then continue moving at a slower pace. Here, I define two generators that yield the expected onscreen deltas for each part of this animation:

```
def move(period, speed):
    for _ in range(period):
        yield speed

def pause(delay):
    for _ in range(delay):
        yield 0
```

To create the final animation, I need to combine move and pause together to produce a single sequence of onscreen deltas. Here, I do this by calling a generator for each step of the animation, iterating over each generator in turn, and then yielding the deltas from all of them in sequence:

```
def animate():
    for delta in move(4, 5.0):
        yield delta
    for delta in pause(3):
        yield delta
    for delta in move(2, 3.0):
        yield delta
```

Now, I can render those deltas onscreen as they're produced by the single animation generator:

```
def render(delta):
    print(f'Delta: {delta:.1f}')
    # Move the images onscreen
    ...

def run(func):
    for delta in func():
        render(delta)

run(animate)

>>>
Delta: 5.0
Delta: 5.0
Delta: 5.0
Delta: 5.0
Delta: 0.0
Delta: 0.0
Delta: 0.0
Delta: 3.0
Delta: 3.0
```

The problem with this code is the repetitive nature of the animate function. The redundancy of the for statements and yield expressions for each generator adds noise and reduces readability. This example includes only three nested generators and it's already hurting clarity; a complex animation with a dozen phases or more would be extremely difficult to follow.

The solution to this problem is to use the yield from expression. This advanced generator feature allows you to yield all values from a nested generator before returning control to the parent generator. Here, I reimplement the animation function by using yield from:

```
def animate_composed():
    yield from move(4, 5.0)
    yield from pause(3)
    yield from move(2, 3.0)

run(animate_composed)
```

```
>>>
Delta: 5.0
Delta: 5.0
Delta: 5.0
Delta: 5.0
Delta: 0.0
Delta: 0.0
Delta: 0.0
Delta: 3.0
Delta: 3.0
```

The result is the same as before, but now the code is clearer and more intuitive. yield from essentially causes the Python interpreter to handle the nested for loop and yield expression boilerplate for you, which results in better performance. Here, I verify the speedup by using the timeit built-in module to run a micro-benchmark:

```
import timeit

def child():
    for i in range(1_000_000):
        yield i

def slow():
    for i in child():
        yield i

def fast():
    yield from child()

baseline = timeit.timeit(
    stmt='for _ in slow(): pass',
    globals=globals(),
    number=50)
print(f'Manual nesting {baseline:.2f}s')
```

```
comparison = timeit.timeit(
    stmt='for _ in fast(): pass',
    globals=globals(),
    number=50)
print(f'Composed nesting {comparison:.2f}s')

reduction = -(comparison - baseline) / baseline
print(f'{reduction:.1%} less time')

>>>
Manual nesting 4.02s
Composed nesting 3.47s
13.5% less time
```

If you find yourself composing generators, I strongly encourage you to use yield from when possible.

Things to Remember

✦ The yield from expression allows you to compose multiple nested generators together into a single combined generator.

✦ yield from provides better performance than manually iterating nested generators and yielding their outputs.

Item 34: Avoid Injecting Data into Generators with send

yield expressions provide generator functions with a simple way to produce an iterable series of output values (see Item 30: "Consider Generators Instead of Returning Lists"). However, this channel appears to be unidirectional: There's no immediately obvious way to simultaneously stream data in and out of a generator as it runs. Having such bidirectional communication could be valuable for a variety of use cases.

For example, say that I'm writing a program to transmit signals using a software-defined radio. Here, I use a function to generate an approximation of a sine wave with a given number of points:

```
import math

def wave(amplitude, steps):
    step_size = 2 * math.pi / steps
    for step in range(steps):
        radians = step * step_size
        fraction = math.sin(radians)
```

```
        output = amplitude * fraction
        yield output
```

Now, I can transmit the wave signal at a single specified amplitude by iterating over the wave generator:

```
def transmit(output):
    if output is None:
        print(f'Output is None')
    else:
        print(f'Output: {output:>5.1f}')


def run(it):
    for output in it:
        transmit(output)


run(wave(3.0, 8))

>>>
Output:   0.0
Output:   2.1
Output:   3.0
Output:   2.1
Output:   0.0
Output:  -2.1
Output:  -3.0
Output:  -2.1
```

This works fine for producing basic waveforms, but it can't be used to constantly vary the amplitude of the wave based on a separate input (i.e., as required to broadcast AM radio signals). I need a way to modulate the amplitude on each iteration of the generator.

Python generators support the send method, which upgrades yield expressions into a two-way channel. The send method can be used to provide streaming inputs to a generator at the same time it's yielding outputs. Normally, when iterating a generator, the value of the yield expression is None:

```
def my_generator():
    received = yield 1
    print(f'received = {received}')


it = iter(my_generator())
output = next(it)        # Get first generator output
print(f'output = {output}')
```

```
try:
    next(it)                # Run generator until it exits
except StopIteration:
    pass

>>>
output = 1
received = None
```

When I call the send method instead of iterating the generator with a for loop or the next built-in function, the supplied parameter becomes the value of the yield expression when the generator is resumed. However, when the generator first starts, a yield expression has not been encountered yet, so the only valid value for calling send initially is None (any other argument would raise an exception at runtime):

```
it = iter(my_generator())
output = it.send(None)  # Get first generator output
print(f'output = {output}')

try:
    it.send('hello!')    # Send value into the generator
except StopIteration:
    pass

>>>
output = 1
received = hello!
```

I can take advantage of this behavior in order to modulate the amplitude of the sine wave based on an input signal. First, I need to change the wave generator to save the amplitude returned by the yield expression and use it to calculate the next generated output:

```
def wave_modulating(steps):
    step_size = 2 * math.pi / steps
    amplitude = yield              # Receive initial amplitude
    for step in range(steps):
        radians = step * step_size
        fraction = math.sin(radians)
        output = amplitude * fraction
        amplitude = yield output  # Receive next amplitude
```

Then, I need to update the run function to stream the modulating amplitude into the wave_modulating generator on each iteration. The first input to send must be None, since a yield expression would not have occurred within the generator yet:

```
def run_modulating(it):
    amplitudes = [
        None, 7, 7, 7, 2, 2, 2, 2, 10, 10, 10, 10, 10]
    for amplitude in amplitudes:
        output = it.send(amplitude)
        transmit(output)

run_modulating(wave_modulating(12))

>>>
Output is None
Output:    0.0
Output:    3.5
Output:    6.1
Output:    2.0
Output:    1.7
Output:    1.0
Output:    0.0
Output:   -5.0
Output:   -8.7
Output:  -10.0
Output:   -8.7
Output:   -5.0
```

This works; it properly varies the output amplitude based on the input signal. The first output is None, as expected, because a value for the amplitude wasn't received by the generator until after the initial yield expression.

One problem with this code is that it's difficult for new readers to understand: Using yield on the right side of an assignment statement isn't intuitive, and it's hard to see the connection between yield and send without already knowing the details of this advanced generator feature.

Now, imagine that the program's requirements get more complicated. Instead of using a simple sine wave as my carrier, I need to use a complex waveform consisting of multiple signals in sequence. One way to implement this behavior is by composing multiple generators together by using the yield from expression (see Item 33: "Compose Multiple Generators with yield from"). Here, I confirm that this works as expected in the simpler case where the amplitude is fixed:

```
def complex_wave():
    yield from wave(7.0, 3)
```

```
    yield from wave(2.0, 4)
    yield from wave(10.0, 5)

run(complex_wave())

>>>
Output:    0.0
Output:    6.1
Output:   -6.1
Output:    0.0
Output:    2.0
Output:    0.0
Output:   -2.0
Output:    0.0
Output:    9.5
Output:    5.9
Output:   -5.9
Output:   -9.5
```

Given that the yield from expression handles the simpler case, you
may expect it to also work properly along with the generator send
method. Here, I try to use it this way by composing multiple calls to
the wave_modulating generator together:

```
def complex_wave_modulating():
    yield from wave_modulating(3)
    yield from wave_modulating(4)
    yield from wave_modulating(5)

run_modulating(complex_wave_modulating())

>>>
Output is None
Output:    0.0
Output:    6.1
Output:   -6.1
Output is None
Output:    0.0
Output:    2.0
Output:    0.0
Output: -10.0
Output is None
Output:    0.0
Output:    9.5
Output:    5.9
```

This works to some extent, but the result contains a big surprise: There are many None values in the output! Why does this happen? When each yield from expression finishes iterating over a nested generator, it moves on to the next one. Each nested generator starts with a bare yield expression—one without a value—in order to receive the initial amplitude from a generator send method call. This causes the parent generator to output a None value when it transitions between child generators.

This means that assumptions about how the yield from and send features behave individually will be broken if you try to use them together. Although it's possible to work around this None problem by increasing the complexity of the run_modulating function, it's not worth the trouble. It's already difficult for new readers of the code to understand how send works. This surprising gotcha with yield from makes it even worse. My advice is to avoid the send method entirely and go with a simpler approach.

The easiest solution is to pass an iterator into the wave function. The iterator should return an input amplitude each time the next built-in function is called on it. This arrangement ensures that each generator is progressed in a cascade as inputs and outputs are processed (see Item 32: "Consider Generator Expressions for Large List Comprehensions" for another example):

```
def wave_cascading(amplitude_it, steps):
    step_size = 2 * math.pi / steps
    for step in range(steps):
        radians = step * step_size
        fraction = math.sin(radians)
        amplitude = next(amplitude_it)  # Get next input
        output = amplitude * fraction
        yield output
```

I can pass the same iterator into each of the generator functions that I'm trying to compose together. Iterators are stateful (see Item 31: "Be Defensive When Iterating Over Arguments"), and thus each of the nested generators picks up where the previous generator left off:

```
def complex_wave_cascading(amplitude_it):
    yield from wave_cascading(amplitude_it, 3)
    yield from wave_cascading(amplitude_it, 4)
    yield from wave_cascading(amplitude_it, 5)
```

Now, I can run the composed generator by simply passing in an iterator from the amplitudes list:

```
def run_cascading():
    amplitudes = [7, 7, 7, 2, 2, 2, 2, 10, 10, 10, 10, 10]
```

```
it = complex_wave_cascading(iter(amplitudes))
for amplitude in amplitudes:
    output = next(it)
    transmit(output)
```

```
run_cascading()
```

```
>>>
Output:    0.0
Output:    6.1
Output:   -6.1
Output:    0.0
Output:    2.0
Output:    0.0
Output:   -2.0
Output:    0.0
Output:    9.5
Output:    5.9
Output:   -5.9
Output:   -9.5
```

The best part about this approach is that the iterator can come from anywhere and could be completely dynamic (e.g., implemented using a generator function). The only downside is that this code assumes that the input generator is completely thread safe, which may not be the case. If you need to cross thread boundaries, async functions may be a better fit (see Item 62: "Mix Threads and Coroutines to Ease the Transition to asyncio").

Things to Remember

✦ The send method can be used to inject data into a generator by giving the yield expression a value that can be assigned to a variable.

✦ Using send with yield from expressions may cause surprising behavior, such as None values appearing at unexpected times in the generator output.

✦ Providing an input iterator to a set of composed generators is a better approach than using the send method, which should be avoided.

Item 35: Avoid Causing State Transitions in Generators with throw

In addition to yield from expressions (see Item 33: "Compose Multiple Generators with yield from") and the send method (see Item 34: "Avoid Injecting Data into Generators with send"), another advanced generator feature is the throw method for re-raising Exception instances within generator functions. The way throw works is simple: When the method is called, the next occurrence of a yield expression re-raises the provided Exception instance after its output is received instead of continuing normally. Here, I show a simple example of this behavior in action:

```
class MyError(Exception):
    pass

def my_generator():
    yield 1
    yield 2
    yield 3

it = my_generator()
print(next(it))  # Yield 1
print(next(it))  # Yield 2
print(it.throw(MyError('test error')))

>>>
1
2
Traceback ...
MyError: test error
```

When you call throw, the generator function may catch the injected exception with a standard try/except compound statement that surrounds the last yield expression that was executed (see Item 65: "Take Advantage of Each Block in try/except/else/finally" for more about exception handling):

```
def my_generator():
    yield 1

    try:
        yield 2
    except MyError:
        print('Got MyError!')
    else:
        yield 3

    yield 4

it = my_generator()
print(next(it))  # Yield 1
```

```
print(next(it))  # Yield 2
print(it.throw(MyError('test error')))

>>>
1
2
Got MyError!
4
```

This functionality provides a two-way communication channel between a generator and its caller that can be useful in certain situations (see Item 34: "Avoid Injecting Data into Generators with send" for another one). For example, imagine that I'm trying to write a program with a timer that supports sporadic resets. Here, I implement this behavior by defining a generator that relies on the throw method:

```
class Reset(Exception):
    pass

def timer(period):
    current = period
    while current:
        current -= 1
        try:
            yield current
        except Reset:
            current = period
```

In this code, whenever the Reset exception is raised by the yield expression, the counter resets itself to its original period.

I can connect this counter reset event to an external input that's polled every second. Then, I can define a run function to drive the timer generator, which injects exceptions with throw to cause resets, or calls announce for each generator output:

```
def check_for_reset():
    # Poll for external event
    ...

def announce(remaining):
    print(f'{remaining} ticks remaining')

def run():
    it = timer(4)
    while True:
```

```
    try:
        if check_for_reset():
            current = it.throw(Reset())
        else:
            current = next(it)
    except StopIteration:
        break
    else:
        announce(current)

run()

>>>
3 ticks remaining
2 ticks remaining
1 ticks remaining
3 ticks remaining
2 ticks remaining
3 ticks remaining
2 ticks remaining
1 ticks remaining
0 ticks remaining
```

This code works as expected, but it's much harder to read than necessary. The various levels of nesting required to catch StopIteration exceptions or decide to throw, call next, or announce make the code noisy.

A simpler approach to implementing this functionality is to define a stateful closure (see Item 38: "Accept Functions Instead of Classes for Simple Interfaces") using an iterable container object (see Item 31: "Be Defensive When Iterating Over Arguments"). Here, I redefine the timer generator by using such a class:

```
class Timer:
    def __init__(self, period):
        self.current = period
        self.period = period

    def reset(self):
        self.current = self.period

    def __iter__(self):
        while self.current:
            self.current -= 1
            yield self.current
```

Now, the run method can do a much simpler iteration by using a for statement, and the code is much easier to follow because of the reduction in the levels of nesting:

```
def run():
    timer = Timer(4)
    for current in timer:
        if check_for_reset():
            timer.reset()
        announce(current)

run()

>>>
3 ticks remaining
2 ticks remaining
1 ticks remaining
3 ticks remaining
2 ticks remaining
3 ticks remaining
2 ticks remaining
1 ticks remaining
0 ticks remaining
```

The output matches the earlier version using throw, but this implementation is much easier to understand, especially for new readers of the code. Often, what you're trying to accomplish by mixing generators and exceptions is better achieved with asynchronous features (see Item 60: "Achieve Highly Concurrent I/O with Coroutines"). Thus, I suggest that you avoid using throw entirely and instead use an iterable class if you need this type of exceptional behavior.

Things to Remember

✦ The throw method can be used to re-raise exceptions within generators at the position of the most recently executed yield expression.

✦ Using throw harms readability because it requires additional nesting and boilerplate in order to raise and catch exceptions.

✦ A better way to provide exceptional behavior in generators is to use a class that implements the __iter__ method along with methods to cause exceptional state transitions.

Item 36: Consider itertools for Working with Iterators and Generators

The itertools built-in module contains a large number of functions that are useful for organizing and interacting with iterators (see Item 30: "Consider Generators Instead of Returning Lists" and Item 31: "Be Defensive When Iterating Over Arguments" for background):

```
import itertools
```

Whenever you find yourself dealing with tricky iteration code, it's worth looking at the itertools documentation again to see if there's anything in there for you to use (see help(itertools)). The following sections describe the most important functions that you should know in three primary categories.

Linking Iterators Together

The itertools built-in module includes a number of functions for linking iterators together.

chain

Use chain to combine multiple iterators into a single sequential iterator:

```
it = itertools.chain([1, 2, 3], [4, 5, 6])
print(list(it))
```

```
>>>
[1, 2, 3, 4, 5, 6]
```

repeat

Use repeat to output a single value forever, or use the second parameter to specify a maximum number of times:

```
it = itertools.repeat('hello', 3)
print(list(it))
```

```
>>>
['hello', 'hello', 'hello']
```

cycle

Use cycle to repeat an iterator's items forever:

```
it = itertools.cycle([1, 2])
result = [next(it) for _ in range (10)]
print(result)
```

```
>>>
[1, 2, 1, 2, 1, 2, 1, 2, 1, 2]
```

tee

Use tee to split a single iterator into the number of parallel iterators specified by the second parameter. The memory usage of this function will grow if the iterators don't progress at the same speed since buffering will be required to enqueue the pending items:

```
it1, it2, it3 = itertools.tee(['first', 'second'], 3)
print(list(it1))
print(list(it2))
print(list(it3))
```

```
>>>
['first', 'second']
['first', 'second']
['first', 'second']
```

zip_longest

This variant of the zip built-in function (see Item 8: "Use zip to Process Iterators in Parallel") returns a placeholder value when an iterator is exhausted, which may happen if iterators have different lengths:

```
keys = ['one', 'two', 'three']
values = [1, 2]

normal = list(zip(keys, values))
print('zip:          ', normal)

it = itertools.zip_longest(keys, values, fillvalue='nope')
longest = list(it)
print('zip_longest:', longest)
```

```
>>>
zip:          [('one', 1), ('two', 2)]
zip_longest: [('one', 1), ('two', 2), ('three', 'nope')]
```

Filtering Items from an Iterator

The `itertools` built-in module includes a number of functions for filtering items from an iterator.

islice

Use islice to slice an iterator by numerical indexes without copying. You can specify the end, start and end, or start, end, and step sizes, and the behavior is similar to that of standard sequence slicing and striding (see Item 11: "Know How to Slice Sequences" and Item 12: "Avoid Striding and Slicing in a Single Expression"):

```
values = [1, 2, 3, 4, 5, 6, 7, 8, 9, 10]

first_five = itertools.islice(values, 5)
print('First five: ', list(first_five))

middle_odds = itertools.islice(values, 2, 8, 2)
print('Middle odds:', list(middle_odds))

>>>
First five:  [1, 2, 3, 4, 5]
Middle odds: [3, 5, 7]
```

takewhile

takewhile returns items from an iterator until a predicate function returns False for an item:

```
values = [1, 2, 3, 4, 5, 6, 7, 8, 9, 10]
less_than_seven = lambda x: x < 7
it = itertools.takewhile(less_than_seven, values)
print(list(it))

>>>
[1, 2, 3, 4, 5, 6]
```

dropwhile

dropwhile, which is the opposite of takewhile, skips items from an iterator until the predicate function returns True for the first time:

```
values = [1, 2, 3, 4, 5, 6, 7, 8, 9, 10]
less_than_seven = lambda x: x < 7
it = itertools.dropwhile(less_than_seven, values)
print(list(it))

>>>
[7, 8, 9, 10]
```

filterfalse

`filterfalse`, which is the opposite of the `filter` built-in function, returns all items from an iterator where a predicate function returns False:

```
values = [1, 2, 3, 4, 5, 6, 7, 8, 9, 10]
evens = lambda x: x % 2 == 0

filter_result = filter(evens, values)
print('Filter:      ', list(filter_result))

filter_false_result = itertools.filterfalse(evens, values)
print('Filter false:', list(filter_false_result))

>>>
Filter:       [2, 4, 6, 8, 10]
Filter false: [1, 3, 5, 7, 9]
```

Producing Combinations of Items from Iterators

The `itertools` built-in module includes a number of functions for producing combinations of items from iterators.

accumulate

`accumulate` folds an item from the iterator into a running value by applying a function that takes two parameters. It outputs the current accumulated result for each input value:

```
values = [1, 2, 3, 4, 5, 6, 7, 8, 9, 10]
sum_reduce = itertools.accumulate(values)
print('Sum:   ', list(sum_reduce))

def sum_modulo_20(first, second):
    output = first + second
    return output % 20

modulo_reduce = itertools.accumulate(values, sum_modulo_20)
print('Modulo:', list(modulo_reduce))

>>>
Sum:    [1, 3, 6, 10, 15, 21, 28, 36, 45, 55]
Modulo: [1, 3, 6, 10, 15, 1, 8, 16, 5, 15]
```

This is essentially the same as the reduce function from the `functools` built-in module, but with outputs yielded one step at a time. By default it sums the inputs if no binary function is specified.

product

product returns the Cartesian product of items from one or more iterators, which is a nice alternative to using deeply nested list comprehensions (see Item 28: "Avoid More Than Two Control Subexpressions in Comprehensions" for why to avoid those):

```
single = itertools.product([1, 2], repeat=2)
print('Single:  ', list(single))

multiple = itertools.product([1, 2], ['a', 'b'])
print('Multiple:', list(multiple))

>>>
Single:   [(1, 1), (1, 2), (2, 1), (2, 2)]
Multiple: [(1, 'a'), (1, 'b'), (2, 'a'), (2, 'b')]
```

permutations

permutations returns the unique ordered permutations of length N with items from an iterator:

```
it = itertools.permutations([1, 2, 3, 4], 2)
print(list(it))

>>>
[(1, 2),
 (1, 3),
 (1, 4),
 (2, 1),
 (2, 3),
 (2, 4),
 (3, 1),
 (3, 2),
 (3, 4),
 (4, 1),
 (4, 2),
 (4, 3)]
```

combinations

combinations returns the unordered combinations of length N with unrepeated items from an iterator:

```
it = itertools.combinations([1, 2, 3, 4], 2)
print(list(it))

>>>
[(1, 2), (1, 3), (1, 4), (2, 3), (2, 4), (3, 4)]
```

combinations_with_replacement

combinations_with_replacement is the same as combinations, but repeated values are allowed:

```
it = itertools.combinations_with_replacement([1, 2, 3, 4], 2)
print(list(it))
```

```
>>>
[(1, 1),
 (1, 2),
 (1, 3),
 (1, 4),
 (2, 2),
 (2, 3),
 (2, 4),
 (3, 3),
 (3, 4),
 (4, 4)]
```

Things to Remember

✦ The itertools functions fall into three main categories for working with iterators and generators: linking iterators together, filtering items they output, and producing combinations of items.

✦ There are more advanced functions, additional parameters, and useful recipes available in the documentation at help(itertools).

Chapter 5

Classes and Interfaces

As an object-oriented programming language, Python supports a full range of features, such as inheritance, polymorphism, and encapsulation. Getting things done in Python often requires writing new classes and defining how they interact through their interfaces and hierarchies.

Python's classes and inheritance make it easy to express a program's intended behaviors with objects. They allow you to improve and expand functionality over time. They provide flexibility in an environment of changing requirements. Knowing how to use them well enables you to write maintainable code.

Item 37: Compose Classes Instead of Nesting Many Levels of Built-in Types

Python's built-in dictionary type is wonderful for maintaining dynamic internal state over the lifetime of an object. By *dynamic*, I mean situations in which you need to do bookkeeping for an unexpected set of identifiers. For example, say that I want to record the grades of a set of students whose names aren't known in advance. I can define a class to store the names in a dictionary instead of using a predefined attribute for each student:

```python
class SimpleGradebook:
    def __init__(self):
        self._grades = {}

    def add_student(self, name):
        self._grades[name] = []

    def report_grade(self, name, score):
        self._grades[name].append(score)
```

```
    def average_grade(self, name):
        grades = self._grades[name]
        return sum(grades) / len(grades)
```

Using the class is simple:

```
book = SimpleGradebook()
book.add_student('Isaac Newton')
book.report_grade('Isaac Newton', 90)
book.report_grade('Isaac Newton', 95)
book.report_grade('Isaac Newton', 85)

print(book.average_grade('Isaac Newton'))
```

```
>>>
90.0
```

Dictionaries and their related built-in types are so easy to use that there's a danger of overextending them to write brittle code. For example, say that I want to extend the SimpleGradebook class to keep a list of grades by subject, not just overall. I can do this by changing the _grades dictionary to map student names (its keys) to yet another dictionary (its values). The innermost dictionary will map subjects (its keys) to a list of grades (its values). Here, I do this by using a defaultdict instance for the inner dictionary to handle missing subjects (see Item 17: "Prefer defaultdict Over setdefault to Handle Missing Items in Internal State" for background):

```
from collections import defaultdict

class BySubjectGradebook:
    def __init__(self):
        self._grades = {}                       # Outer dict

    def add_student(self, name):
        self._grades[name] = defaultdict(list)  # Inner dict
```

This seems straightforward enough. The report_grade and average_ grade methods gain quite a bit of complexity to deal with the multilevel dictionary, but it's seemingly manageable:

```
    def report_grade(self, name, subject, grade):
        by_subject = self._grades[name]
        grade_list = by_subject[subject]
        grade_list.append(grade)

    def average_grade(self, name):
        by_subject = self._grades[name]
```

```
        total, count = 0, 0
        for grades in by_subject.values():
            total += sum(grades)
            count += len(grades)
        return total / count
```

Using the class remains simple:

```
book = BySubjectGradebook()
book.add_student('Albert Einstein')
book.report_grade('Albert Einstein', 'Math', 75)
book.report_grade('Albert Einstein', 'Math', 65)
book.report_grade('Albert Einstein', 'Gym', 90)
book.report_grade('Albert Einstein', 'Gym', 95)
print(book.average_grade('Albert Einstein'))
```

```
>>>
81.25
```

Now, imagine that the requirements change again. I also want to track the weight of each score toward the overall grade in the class so that midterm and final exams are more important than pop quizzes. One way to implement this feature is to change the innermost dictionary; instead of mapping subjects (its keys) to a list of grades (its values), I can use the tuple of (score, weight) in the values list:

```
class WeightedGradebook:
    def __init__(self):
        self._grades = {}

    def add_student(self, name):
        self._grades[name] = defaultdict(list)

    def report_grade(self, name, subject, score, weight):
        by_subject = self._grades[name]
        grade_list = by_subject[subject]
        grade_list.append((score, weight))
```

Although the changes to report_grade seem simple—just make the grade list store tuple instances—the average_grade method now has a loop within a loop and is difficult to read:

```
    def average_grade(self, name):
        by_subject = self._grades[name]

        score_sum, score_count = 0, 0
        for subject, scores in by_subject.items():
            subject_avg, total_weight = 0, 0
```

```
        for score, weight in scores:
            subject_avg += score * weight
            total_weight += weight

        score_sum += subject_avg / total_weight
        score_count += 1

    return score_sum / score_count
```

Using the class has also gotten more difficult. It's unclear what all of the numbers in the positional arguments mean:

```
book = WeightedGradebook()
book.add_student('Albert Einstein')
book.report_grade('Albert Einstein', 'Math', 75, 0.05)
book.report_grade('Albert Einstein', 'Math', 65, 0.15)
book.report_grade('Albert Einstein', 'Math', 70, 0.80)
book.report_grade('Albert Einstein', 'Gym', 100, 0.40)
book.report_grade('Albert Einstein', 'Gym', 85, 0.60)
print(book.average_grade('Albert Einstein'))

>>>
80.25
```

When you see complexity like this, it's time to make the leap from built-in types like dictionaries, tuples, sets, and lists to a hierarchy of classes.

In the grades example, at first I didn't know I'd need to support weighted grades, so the complexity of creating classes seemed unwarranted. Python's built-in dictionary and tuple types made it easy to keep going, adding layer after layer to the internal bookkeeping. But you should avoid doing this for more than one level of nesting; using dictionaries that contain dictionaries makes your code hard to read by other programmers and sets you up for a maintenance nightmare.

As soon as you realize that your bookkeeping is getting complicated, break it all out into classes. You can then provide well-defined interfaces that better encapsulate your data. This approach also enables you to create a layer of abstraction between your interfaces and your concrete implementations.

Refactoring to Classes

There are many approaches to refactoring (see Item 89: "Consider warnings to Refactor and Migrate Usage" for another). In this case,

I can start moving to classes at the bottom of the dependency tree: a single grade. A class seems too heavyweight for such simple information. A tuple, though, seems appropriate because grades are immutable. Here, I use the tuple of (score, weight) to track grades in a list:

```
grades = []
grades.append((95, 0.45))
grades.append((85, 0.55))
total = sum(score * weight for score, weight in grades)
total_weight = sum(weight for _, weight in grades)
average_grade = total / total_weight
```

I used _ (the underscore variable name, a Python convention for unused variables) to capture the first entry in each grade's tuple and ignore it when calculating the total_weight.

The problem with this code is that tuple instances are positional. For example, if I want to associate more information with a grade, such as a set of notes from the teacher, I need to rewrite every usage of the two-tuple to be aware that there are now three items present instead of two, which means I need to use _ further to ignore certain indexes:

```
grades = []
grades.append((95, 0.45, 'Great job'))
grades.append((85, 0.55, 'Better next time'))
total = sum(score * weight for score, weight, _ in grades)
total_weight = sum(weight for _, weight, _ in grades)
average_grade = total / total_weight
```

This pattern of extending tuples longer and longer is similar to deepening layers of dictionaries. As soon as you find yourself going longer than a two-tuple, it's time to consider another approach.

The namedtuple type in the collections built-in module does exactly what I need in this case: It lets me easily define tiny, immutable data classes:

```
from collections import namedtuple


Grade = namedtuple('Grade', ('score', 'weight'))
```

These classes can be constructed with positional or keyword arguments. The fields are accessible with named attributes. Having named attributes makes it easy to move from a namedtuple to a class later if the requirements change again and I need to, say, support mutability or behaviors in the simple data containers.

> ## Limitations of namedtuple
>
> Although namedtuple is useful in many circumstances, it's import-
> ant to understand when it can do more harm than good:
>
> - You can't specify default argument values for namedtuple
>   classes. This makes them unwieldy when your data may have
>   many optional properties. If you find yourself using more than
>   a handful of attributes, using the built-in dataclasses module
>   may be a better choice.
>
> - The attribute values of namedtuple instances are still accessi-
>   ble using numerical indexes and iteration. Especially in exter-
>   nalized APIs, this can lead to unintentional usage that makes
>   it harder to move to a real class later. If you're not in control
>   of all of the usage of your namedtuple instances, it's better to
>   explicitly define a new class.

Next, I can write a class to represent a single subject that contains a
set of grades:

```python
class Subject:
    def __init__(self):
        self._grades = []

    def report_grade(self, score, weight):
        self._grades.append(Grade(score, weight))

    def average_grade(self):
        total, total_weight = 0, 0
        for grade in self._grades:
            total += grade.score * grade.weight
            total_weight += grade.weight
        return total / total_weight
```

Then, I write a class to represent a set of subjects that are being stud-
ied by a single student:

```python
class Student:
    def __init__(self):
        self._subjects = defaultdict(Subject)

    def get_subject(self, name):
        return self._subjects[name]
```

```
def average_grade(self):
    total, count = 0, 0
    for subject in self._subjects.values():
        total += subject.average_grade()
        count += 1
    return total / count
```

Finally, I'd write a container for all of the students, keyed dynamically by their names:

```
class Gradebook:
    def __init__(self):
        self._students = defaultdict(Student)

    def get_student(self, name):
        return self._students[name]
```

The line count of these classes is almost double the previous implementation's size. But this code is much easier to read. The example driving the classes is also more clear and extensible:

```
book = Gradebook()
albert = book.get_student('Albert Einstein')
math = albert.get_subject('Math')
math.report_grade(75, 0.05)
math.report_grade(65, 0.15)
math.report_grade(70, 0.80)
gym = albert.get_subject('Gym')
gym.report_grade(100, 0.40)
gym.report_grade(85, 0.60)
print(albert.average_grade())
```

```
>>>
80.25
```

It would also be possible to write backward-compatible methods to help migrate usage of the old API style to the new hierarchy of objects.

Things to Remember

✦ Avoid making dictionaries with values that are dictionaries, long tuples, or complex nestings of other built-in types.

✦ Use namedtuple for lightweight, immutable data containers before you need the flexibility of a full class.

✦ Move your bookkeeping code to using multiple classes when your internal state dictionaries get complicated.

Item 38: Accept Functions Instead of Classes for Simple Interfaces

Many of Python's built-in APIs allow you to customize behavior by passing in a function. These *hooks* are used by APIs to call back your code while they execute. For example, the list type's sort method takes an optional key argument that's used to determine each index's value for sorting (see Item 14: "Sort by Complex Criteria Using the key Parameter" for details). Here, I sort a list of names based on their lengths by providing the len built-in function as the key hook:

```
names = ['Socrates', 'Archimedes', 'Plato', 'Aristotle']
names.sort(key=len)
print(names)
```

```
>>>
['Plato', 'Socrates', 'Aristotle', 'Archimedes']
```

In other languages, you might expect hooks to be defined by an abstract class. In Python, many hooks are just stateless functions with well-defined arguments and return values. Functions are ideal for hooks because they are easier to describe and simpler to define than classes. Functions work as hooks because Python has *first-class* functions: Functions and methods can be passed around and referenced like any other value in the language.

For example, say that I want to customize the behavior of the defaultdict class (see Item 17: "Prefer defaultdict Over setdefault to Handle Missing Items in Internal State" for background). This data structure allows you to supply a function that will be called with no arguments each time a missing key is accessed. The function must return the default value that the missing key should have in the dictionary. Here, I define a hook that logs each time a key is missing and returns 0 for the default value:

```
def log_missing():
    print('Key added')
    return 0
```

Given an initial dictionary and a set of desired increments, I can cause the log_missing function to run and print twice (for 'red' and 'orange'):

```
from collections import defaultdict

current = {'green': 12, 'blue': 3}
increments = [
```

```
        ('red', 5),
        ('blue', 17),
        ('orange', 9),
    ]
    result = defaultdict(log_missing, current)
    print('Before:', dict(result))
    for key, amount in increments:
        result[key] += amount
    print('After: ', dict(result))

    >>>
    Before: {'green': 12, 'blue': 3}
    Key added
    Key added
    After:  {'green': 12, 'blue': 20, 'red': 5, 'orange': 9}
```

Supplying functions like log_missing makes APIs easy to build and test because it separates side effects from deterministic behavior. For example, say I now want the default value hook passed to defaultdict to count the total number of keys that were missing. One way to achieve this is by using a stateful closure (see Item 21: "Know How Closures Interact with Variable Scope" for details). Here, I define a helper function that uses such a closure as the default value hook:

```
def increment_with_report(current, increments):
    added_count = 0

    def missing():
        nonlocal added_count  # Stateful closure
        added_count += 1
        return 0

    result = defaultdict(missing, current)
    for key, amount in increments:
        result[key] += amount

    return result, added_count
```

Running this function produces the expected result (2), even though the defaultdict has no idea that the missing hook maintains state. Another benefit of accepting simple functions for interfaces is that it's easy to add functionality later by hiding state in a closure:

```
result, count = increment_with_report(current, increments)
assert count == 2
```

The problem with defining a closure for stateful hooks is that it's harder to read than the stateless function example. Another approach is to define a small class that encapsulates the state you want to track:

```
class CountMissing:
    def __init__(self):
        self.added = 0

    def missing(self):
        self.added += 1
        return 0
```

In other languages, you might expect that now defaultdict would have to be modified to accommodate the interface of CountMissing. But in Python, thanks to first-class functions, you can reference the CountMissing.missing method directly on an object and pass it to defaultdict as the default value hook. It's trivial to have an object instance's method satisfy a function interface:

```
counter = CountMissing()
result = defaultdict(counter.missing, current)  # Method ref
for key, amount in increments:
    result[key] += amount
assert counter.added == 2
```

Using a helper class like this to provide the behavior of a stateful closure is clearer than using the increment_with_report function, as above. However, in isolation, it's still not immediately obvious what the purpose of the CountMissing class is. Who constructs a CountMissing object? Who calls the missing method? Will the class need other public methods to be added in the future? Until you see its usage with defaultdict, the class is a mystery.

To clarify this situation, Python allows classes to define the __call__ special method. __call__ allows an object to be called just like a function. It also causes the callable built-in function to return True for such an instance, just like a normal function or method. All objects that can be executed in this manner are referred to as *callables*:

```
class BetterCountMissing:
    def __init__(self):
        self.added = 0

    def __call__(self):
        self.added += 1
        return 0
```

```
counter = BetterCountMissing()
assert counter() == 0
assert callable(counter)
```

Here, I use a BetterCountMissing instance as the default value hook for a defaultdict to track the number of missing keys that were added:

```
counter = BetterCountMissing()
result = defaultdict(counter, current)  # Relies on __call__
for key, amount in increments:
    result[key] += amount
assert counter.added == 2
```

This is much clearer than the CountMissing.missing example. The __call__ method indicates that a class's instances will be used somewhere a function argument would also be suitable (like API hooks). It directs new readers of the code to the entry point that's responsible for the class's primary behavior. It provides a strong hint that the goal of the class is to act as a stateful closure.

Best of all, defaultdict still has no view into what's going on when you use __call__. All that defaultdict requires is a function for the default value hook. Python provides many different ways to satisfy a simple function interface, and you can choose the one that works best for what you need to accomplish.

Things to Remember

+ Instead of defining and instantiating classes, you can often simply use functions for simple interfaces between components in Python.

+ References to functions and methods in Python are first class, meaning they can be used in expressions (like any other type).

+ The __call__ special method enables instances of a class to be called like plain Python functions.

+ When you need a function to maintain state, consider defining a class that provides the __call__ method instead of defining a stateful closure.

Item 39: Use @classmethod Polymorphism to Construct Objects Generically

In Python, not only do objects support polymorphism, but classes do as well. What does that mean, and what is it good for?

Polymorphism enables multiple classes in a hierarchy to implement their own unique versions of a method. This means that many classes can fulfill the same interface or abstract base class while providing different functionality (see Item 43: "Inherit from collections.abc for Custom Container Types").

For example, say that I'm writing a MapReduce implementation, and I want a common class to represent the input data. Here, I define such a class with a read method that must be defined by subclasses:

```
class InputData:
    def read(self):
        raise NotImplementedError
```

I also have a concrete subclass of InputData that reads data from a file on disk:

```
class PathInputData(InputData):
    def __init__(self, path):
        super().__init__()
        self.path = path

    def read(self):
        with open(self.path) as f:
            return f.read()
```

I could have any number of InputData subclasses, like PathInputData, and each of them could implement the standard interface for read to return the data to process. Other InputData subclasses could read from the network, decompress data transparently, and so on.

I'd want a similar abstract interface for the MapReduce worker that consumes the input data in a standard way:

```
class Worker:
    def __init__(self, input_data):
        self.input_data = input_data
        self.result = None

    def map(self):
        raise NotImplementedError

    def reduce(self, other):
        raise NotImplementedError
```

Here, I define a concrete subclass of Worker to implement the specific MapReduce function I want to apply—a simple newline counter:

```
class LineCountWorker(Worker):
    def map(self):
        data = self.input_data.read()
        self.result = data.count('\n')
```

```
def reduce(self, other):
    self.result += other.result
```

It may look like this implementation is going great, but I've reached the biggest hurdle in all of this. What connects all of these pieces? I have a nice set of classes with reasonable interfaces and abstractions, but that's only useful once the objects are constructed. What's responsible for building the objects and orchestrating the MapReduce?

The simplest approach is to manually build and connect the objects with some helper functions. Here, I list the contents of a directory and construct a PathInputData instance for each file it contains:

```
import os


def generate_inputs(data_dir):
    for name in os.listdir(data_dir):
        yield PathInputData(os.path.join(data_dir, name))
```

Next, I create the LineCountWorker instances by using the InputData instances returned by generate_inputs:

```
def create_workers(input_list):
    workers = []
    for input_data in input_list:
        workers.append(LineCountWorker(input_data))
    return workers
```

I execute these Worker instances by fanning out the map step to multiple threads (see Item 53: "Use Threads for Blocking I/O, Avoid for Parallelism" for background). Then, I call reduce repeatedly to combine the results into one final value:

```
from threading import Thread


def execute(workers):
    threads = [Thread(target=w.map) for w in workers]
    for thread in threads: thread.start()
    for thread in threads: thread.join()

    first, *rest = workers
    for worker in rest:
        first.reduce(worker)
    return first.result
```

Finally, I connect all the pieces together in a function to run each step:

```
def mapreduce(data_dir):
    inputs = generate_inputs(data_dir)
    workers = create_workers(inputs)
    return execute(workers)
```

Running this function on a set of test input files works great:

```
import os
import random

def write_test_files(tmpdir):
    os.makedirs(tmpdir)
    for i in range(100):
        with open(os.path.join(tmpdir, str(i)), 'w') as f:
            f.write('\n' * random.randint(0, 100))

tmpdir = 'test_inputs'
write_test_files(tmpdir)

result = mapreduce(tmpdir)
print(f'There are {result} lines')

>>>
There are 4360 lines
```

What's the problem? The huge issue is that the mapreduce function is not generic at all. If I wanted to write another InputData or Worker subclass, I would also have to rewrite the generate_inputs, create_workers, and mapreduce functions to match.

This problem boils down to needing a generic way to construct objects. In other languages, you'd solve this problem with constructor polymorphism, requiring that each InputData subclass provides a special constructor that can be used generically by the helper methods that orchestrate the MapReduce (similar to the factory pattern). The trouble is that Python only allows for the single constructor method __init__. It's unreasonable to require every InputData subclass to have a compatible constructor.

The best way to solve this problem is with *class method* polymorphism. This is exactly like the instance method polymorphism I used for InputData.read, except that it's for whole classes instead of their constructed objects.

Let me apply this idea to the MapReduce classes. Here, I extend the
InputData class with a generic @classmethod that's responsible for creating new InputData instances using a common interface:

```
class GenericInputData:
    def read(self):
        raise NotImplementedError

    @classmethod
    def generate_inputs(cls, config):
        raise NotImplementedError
```

I have generate_inputs take a dictionary with a set of configuration
parameters that the GenericInputData concrete subclass needs to interpret. Here, I use the config to find the directory to list for input files:

```
class PathInputData(GenericInputData):
    ...

    @classmethod
    def generate_inputs(cls, config):
        data_dir = config['data_dir']
        for name in os.listdir(data_dir):
            yield cls(os.path.join(data_dir, name))
```

Similarly, I can make the create_workers helper part of the
GenericWorker class. Here, I use the input_class parameter, which
must be a subclass of GenericInputData, to generate the necessary
inputs. I construct instances of the GenericWorker concrete subclass
by using cls() as a generic constructor:

```
class GenericWorker:
    def __init__(self, input_data):
        self.input_data = input_data
        self.result = None

    def map(self):
        raise NotImplementedError

    def reduce(self, other):
        raise NotImplementedError

    @classmethod
    def create_workers(cls, input_class, config):
        workers = []
        for input_data in input_class.generate_inputs(config):
            workers.append(cls(input_data))
        return workers
```

Note that the call to input_class.generate_inputs above is the class polymorphism that I'm trying to show. You can also see how create_workers calling cls() provides an alternative way to construct GenericWorker objects besides using the __init__ method directly.

The effect on my concrete GenericWorker subclass is nothing more than changing its parent class:

```
class LineCountWorker(GenericWorker):
    ...
```

Finally, I can rewrite the mapreduce function to be completely generic by calling create_workers:

```
def mapreduce(worker_class, input_class, config):
    workers = worker_class.create_workers(input_class, config)
    return execute(workers)
```

Running the new worker on a set of test files produces the same result as the old implementation. The difference is that the mapreduce function requires more parameters so that it can operate generically:

```
config = {'data_dir': tmpdir}
result = mapreduce(LineCountWorker, PathInputData, config)
print(f'There are {result} lines')

>>>
There are 4360 lines
```

Now, I can write other GenericInputData and GenericWorker subclasses as I wish, without having to rewrite any of the glue code.

Things to Remember

✦ Python only supports a single constructor per class: the __init__ method.

✦ Use @classmethod to define alternative constructors for your classes.

✦ Use class method polymorphism to provide generic ways to build and connect many concrete subclasses.

Item 40: Initialize Parent Classes with super

The old, simple way to initialize a parent class from a child class is to directly call the parent class's __init__ method with the child instance:

```
class MyBaseClass:
    def __init__(self, value):
        self.value = value
```

```
class MyChildClass(MyBaseClass):
    def __init__(self):
        MyBaseClass.__init__(self, 5)
```

This approach works fine for basic class hierarchies but breaks in many cases.

If a class is affected by multiple inheritance (something to avoid in general; see Item 41: "Consider Composing Functionality with Mix-in Classes"), calling the superclasses' __init__ methods directly can lead to unpredictable behavior.

One problem is that the __init__ call order isn't specified across all subclasses. For example, here I define two parent classes that operate on the instance's value field:

```
class TimesTwo:
    def __init__(self):
        self.value *= 2


class PlusFive:
    def __init__(self):
        self.value += 5
```

This class defines its parent classes in one ordering:

```
class OneWay(MyBaseClass, TimesTwo, PlusFive):
    def __init__(self, value):
        MyBaseClass.__init__(self, value)
        TimesTwo.__init__(self)
        PlusFive.__init__(self)
```

And constructing it produces a result that matches the parent class ordering:

```
foo = OneWay(5)
print('First ordering value is (5 * 2) + 5 =', foo.value)

>>>
First ordering value is (5 * 2) + 5 = 15
```

Here's another class that defines the same parent classes but in a different ordering (PlusFive followed by TimesTwo instead of the other way around):

```
class AnotherWay(MyBaseClass, PlusFive, TimesTwo):
    def __init__(self, value):
        MyBaseClass.__init__(self, value)
        TimesTwo.__init__(self)
        PlusFive.__init__(self)
```

However, I left the calls to the parent class constructors— PlusFive.__init__ and TimesTwo.__init__—in the same order as before, which means this class's behavior doesn't match the order of the parent classes in its definition. The conflict here between the inheritance base classes and the __init__ calls is hard to spot, which makes this especially difficult for new readers of the code to understand:

```
bar = AnotherWay(5)
print('Second ordering value is', bar.value)
```

```
>>>
Second ordering value is 15
```

Another problem occurs with diamond inheritance. Diamond inheritance happens when a subclass inherits from two separate classes that have the same superclass somewhere in the hierarchy. Diamond inheritance causes the common superclass's __init__ method to run multiple times, causing unexpected behavior. For example, here I define two child classes that inherit from MyBaseClass:

```
class TimesSeven(MyBaseClass):
    def __init__(self, value):
        MyBaseClass.__init__(self, value)
        self.value *= 7
```

```
class PlusNine(MyBaseClass):
    def __init__(self, value):
        MyBaseClass.__init__(self, value)
        self.value += 9
```

Then, I define a child class that inherits from both of these classes, making MyBaseClass the top of the diamond:

```
class ThisWay(TimesSeven, PlusNine):
    def __init__(self, value):
        TimesSeven.__init__(self, value)
        PlusNine.__init__(self, value)
```

```
foo = ThisWay(5)
print('Should be (5 * 7) + 9 = 44 but is', foo.value)
```

```
>>>
Should be (5 * 7) + 9 = 44 but is 14
```

The call to the second parent class's constructor, PlusNine.__init__, causes self.value to be reset back to 5 when MyBaseClass.__init__ gets called a second time. That results in the calculation of self.value to be 5 + 9 = 14, completely ignoring the effect of the TimesSeven.__init__ constructor. This behavior is surprising and can be very difficult to debug in more complex cases.

To solve these problems, Python has the super built-in function and standard method resolution order (MRO). super ensures that common superclasses in diamond hierarchies are run only once (for another example, see Item 48: "Validate Subclasses with __init_subclass__"). The MRO defines the ordering in which superclasses are initialized, following an algorithm called *C3 linearization*.

Here, I create a diamond-shaped class hierarchy again, but this time I use super to initialize the parent class:

```python
class TimesSevenCorrect(MyBaseClass):
    def __init__(self, value):
        super().__init__(value)
        self.value *= 7


class PlusNineCorrect(MyBaseClass):
    def __init__(self, value):
        super().__init__(value)
        self.value += 9
```

Now, the top part of the diamond, MyBaseClass.__init__, is run only a single time. The other parent classes are run in the order specified in the class statement:

```python
class GoodWay(TimesSevenCorrect, PlusNineCorrect):
    def __init__(self, value):
        super().__init__(value)


foo = GoodWay(5)
print('Should be 7 * (5 + 9) = 98 and is', foo.value)

>>>
Should be 7 * (5 + 9) = 98 and is 98
```

This order may seem backward at first. Shouldn't TimesSevenCorrect. __init__ have run first? Shouldn't the result be (5 * 7) + 9 = 44? The answer is no. This ordering matches what the MRO defines for this class. The MRO ordering is available on a class method called mro:

```python
mro_str = '\n'.join(repr(cls) for cls in GoodWay.mro())
print(mro_str)

>>>
<class '__main__.GoodWay'>
<class '__main__.TimesSevenCorrect'>
```

```
<class '__main__.PlusNineCorrect'>
<class '__main__.MyBaseClass'>
<class 'object'>
```

When I call GoodWay(5), it in turn calls TimesSevenCorrect.__init__, which calls PlusNineCorrect.__init__, which calls MyBaseClass.__init__. Once this reaches the top of the diamond, all of the initialization methods actually do their work in the opposite order from how their __init__ functions were called. MyBaseClass.__init__ assigns value to 5. PlusNineCorrect.__init__ adds 9 to make value equal 14. TimesSevenCorrect.__init__ multiplies it by 7 to make value equal 98.

Besides making multiple inheritance robust, the call to super().__init__ is also much more maintainable than calling MyBaseClass.__init__ directly from within the subclasses. I could later rename MyBaseClass to something else or have TimesSevenCorrect and PlusNineCorrect inherit from another superclass without having to update their __init__ methods to match.

The super function can also be called with two parameters: first the type of the class whose MRO parent view you're trying to access, and then the instance on which to access that view. Using these optional parameters within the constructor looks like this:

```
class ExplicitTrisect(MyBaseClass):
    def __init__(self, value):
        super(ExplicitTrisect, self).__init__(value)
        self.value /= 3
```

However, these parameters are not required for object instance initialization. Python's compiler automatically provides the correct parameters (__class__ and self) for you when super is called with zero arguments within a class definition. This means all three of these usages are equivalent:

```
class AutomaticTrisect(MyBaseClass):
    def __init__(self, value):
        super(__class__, self).__init__(value)
        self.value /= 3

class ImplicitTrisect(MyBaseClass):
    def __init__(self, value):
        super().__init__(value)
        self.value /= 3

assert ExplicitTrisect(9).value == 3
assert AutomaticTrisect(9).value == 3
assert ImplicitTrisect(9).value == 3
```

The only time you should provide parameters to super is in situations where you need to access the specific functionality of a superclass's implementation from a child class (e.g., to wrap or reuse functionality).

Things to Remember

✦ Python's standard method resolution order (MRO) solves the problems of superclass initialization order and diamond inheritance.

✦ Use the super built-in function with zero arguments to initialize parent classes.

Item 41: Consider Composing Functionality with Mix-in Classes

Python is an object-oriented language with built-in facilities for making multiple inheritance tractable (see Item 40: "Initialize Parent Classes with super"). However, it's better to avoid multiple inheritance altogether.

If you find yourself desiring the convenience and encapsulation that come with multiple inheritance, but want to avoid the potential headaches, consider writing a *mix-in* instead. A mix-in is a class that defines only a small set of additional methods for its child classes to provide. Mix-in classes don't define their own instance attributes nor require their __init__ constructor to be called.

Writing mix-ins is easy because Python makes it trivial to inspect the current state of any object, regardless of its type. Dynamic inspection means you can write generic functionality just once, in a mix-in, and it can then be applied to many other classes. Mix-ins can be composed and layered to minimize repetitive code and maximize reuse.

For example, say I want the ability to convert a Python object from its in-memory representation to a dictionary that's ready for serialization. Why not write this functionality generically so I can use it with all my classes?

Here, I define an example mix-in that accomplishes this with a new public method that's added to any class that inherits from it:

```
class ToDictMixin:
    def to_dict(self):
        return self._traverse_dict(self.__dict__)
```

The implementation details are straightforward and rely on dynamic attribute access using hasattr, dynamic type inspection with isinstance, and accessing the instance dictionary __dict__:

```
def _traverse_dict(self, instance_dict):
    output = {}
    for key, value in instance_dict.items():
        output[key] = self._traverse(key, value)
    return output

def _traverse(self, key, value):
    if isinstance(value, ToDictMixin):
        return value.to_dict()
    elif isinstance(value, dict):
        return self._traverse_dict(value)
    elif isinstance(value, list):
        return [self._traverse(key, i) for i in value]
    elif hasattr(value, '__dict__'):
        return self._traverse_dict(value.__dict__)
    else:
        return value
```

Here, I define an example class that uses the mix-in to make a dictionary representation of a binary tree:

```
class BinaryTree(ToDictMixin):
    def __init__(self, value, left=None, right=None):
        self.value = value
        self.left = left
        self.right = right
```

Translating a large number of related Python objects into a dictionary becomes easy:

```
tree = BinaryTree(10,
    left=BinaryTree(7, right=BinaryTree(9)),
    right=BinaryTree(13, left=BinaryTree(11)))
print(tree.to_dict())

>>>
{'value': 10,
 'left': {'value': 7,
          'left': None,
          'right': {'value': 9, 'left': None, 'right': None}},
 'right': {'value': 13,
           'left': {'value': 11, 'left': None, 'right': None},
           'right': None}}
```

The best part about mix-ins is that you can make their generic functionality pluggable so behaviors can be overridden when required. For example, here I define a subclass of BinaryTree that holds a reference to its parent. This circular reference would cause the default implementation of ToDictMixin.to_dict to loop forever:

```
class BinaryTreeWithParent(BinaryTree):
    def __init__(self, value, left=None,
                 right=None, parent=None):
        super().__init__(value, left=left, right=right)
        self.parent = parent
```

The solution is to override the BinaryTreeWithParent._traverse method to only process values that matter, preventing cycles encountered by the mix-in. Here, the _traverse override inserts the parent's numerical value and otherwise defers to the mix-in's default implementation by using the super built-in function:

```
    def _traverse(self, key, value):
        if (isinstance(value, BinaryTreeWithParent) and
                key == 'parent'):
            return value.value  # Prevent cycles
        else:
            return super()._traverse(key, value)
```

Calling BinaryTreeWithParent.to_dict works without issue because the circular referencing properties aren't followed:

```
root = BinaryTreeWithParent(10)
root.left = BinaryTreeWithParent(7, parent=root)
root.left.right = BinaryTreeWithParent(9, parent=root.left)
print(root.to_dict())

>>>
{'value': 10,
 'left': {'value': 7,
          'left': None,
          'right': {'value': 9,
                    'left': None,
                    'right': None,
                    'parent': 7},
          'parent': 10},
 'right': None,
 'parent': None}
```

By defining BinaryTreeWithParent._traverse, I've also enabled any class that has an attribute of type BinaryTreeWithParent to automatically work with the ToDictMixin:

```
class NamedSubTree(ToDictMixin):
    def __init__(self, name, tree_with_parent):
        self.name = name
        self.tree_with_parent = tree_with_parent

my_tree = NamedSubTree('foobar', root.left.right)
print(my_tree.to_dict())  # No infinite loop
```

```
>>>
{'name': 'foobar',
 'tree_with_parent': {'value': 9,
                      'left': None,
                      'right': None,
                      'parent': 7}}
```

Mix-ins can also be composed together. For example, say I want a mix-in that provides generic JSON serialization for any class. I can do this by assuming that a class provides a to_dict method (which may or may not be provided by the ToDictMixin class):

```
import json


class JsonMixin:
    @classmethod
    def from_json(cls, data):
        kwargs = json.loads(data)
        return cls(**kwargs)

    def to_json(self):
        return json.dumps(self.to_dict())
```

Note how the JsonMixin class defines both instance methods and class methods. Mix-ins let you add either kind of behavior to subclasses. In this example, the only requirements of a JsonMixin subclass are providing a to_dict method and taking keyword arguments for the __init__ method (see Item 23: "Provide Optional Behavior with Keyword Arguments" for background).

This mix-in makes it simple to create hierarchies of utility classes that can be serialized to and from JSON with little boilerplate. For example, here I have a hierarchy of data classes representing parts of a datacenter topology:

```
class DatacenterRack(ToDictMixin, JsonMixin):
    def __init__(self, switch=None, machines=None):
        self.switch = Switch(**switch)
        self.machines = [
            Machine(**kwargs) for kwargs in machines]

class Switch(ToDictMixin, JsonMixin):
    def __init__(self, ports=None, speed=None):
        self.ports = ports
        self.speed = speed

class Machine(ToDictMixin, JsonMixin):
    def __init__(self, cores=None, ram=None, disk=None):
        self.cores = cores
        self.ram = ram
        self.disk = disk
```

Serializing these classes to and from JSON is simple. Here, I verify that the data is able to be sent round-trip through serializing and deserializing:

```
serialized = """{
    "switch": {"ports": 5, "speed": 1e9},
    "machines": [
        {"cores": 8, "ram": 32e9, "disk": 5e12},
        {"cores": 4, "ram": 16e9, "disk": 1e12},
        {"cores": 2, "ram": 4e9, "disk": 500e9}
    ]
}"""

deserialized = DatacenterRack.from_json(serialized)
roundtrip = deserialized.to_json()
assert json.loads(serialized) == json.loads(roundtrip)
```

When you use mix-ins like this, it's fine if the class you apply JsonMixin to already inherits from JsonMixin higher up in the class hierarchy. The resulting class will behave the same way, thanks to the behavior of super.

Things to Remember

♦ Avoid using multiple inheritance with instance attributes and __init__ if mix-in classes can achieve the same outcome.

♦ Use pluggable behaviors at the instance level to provide per-class customization when mix-in classes may require it.

✦ Mix-ins can include instance methods or class methods, depending on your needs.

✦ Compose mix-ins to create complex functionality from simple behaviors.

Item 42: Prefer Public Attributes Over Private Ones

In Python, there are only two types of visibility for a class's attributes: *public* and *private*:

```python
class MyObject:
    def __init__(self):
        self.public_field = 5
        self.__private_field = 10

    def get_private_field(self):
        return self.__private_field
```

Public attributes can be accessed by anyone using the dot operator on the object:

```python
foo = MyObject()
assert foo.public_field == 5
```

Private fields are specified by prefixing an attribute's name with a double underscore. They can be accessed directly by methods of the containing class:

```python
assert foo.get_private_field() == 10
```

However, directly accessing private fields from outside the class raises an exception:

```python
foo.__private_field
```

```
>>>
Traceback ...
AttributeError: 'MyObject' object has no attribute
➥'__private_field'
```

Class methods also have access to private attributes because they are declared within the surrounding class block:

```python
class MyOtherObject:
    def __init__(self):
        self.__private_field = 71

    @classmethod
    def get_private_field_of_instance(cls, instance):
        return instance.__private_field
```

```
bar = MyOtherObject()
assert MyOtherObject.get_private_field_of_instance(bar) == 71
```

As you'd expect with private fields, a subclass can't access its parent class's private fields:

```
class MyParentObject:
    def __init__(self):
        self.__private_field = 71


class MyChildObject(MyParentObject):
    def get_private_field(self):
        return self.__private_field


baz = MyChildObject()
baz.get_private_field()

>>>
Traceback ...
AttributeError: 'MyChildObject' object has no attribute
➥'_MyChildObject__private_field'
```

The private attribute behavior is implemented with a simple transformation of the attribute name. When the Python compiler sees private attribute access in methods like MyChildObject.get_private_field, it translates the __private_field attribute access to use the name _MyChildObject__private_field instead. In the example above, __private_field is only defined in MyParentObject.__init__, which means the private attribute's real name is _MyParentObject__private_field. Accessing the parent's private attribute from the child class fails simply because the transformed attribute name doesn't exist (_MyChildObject__private_field instead of _MyParentObject__private_field).

Knowing this scheme, you can easily access the private attributes of any class—from a subclass or externally—without asking for permission:

```
assert baz._MyParentObject__private_field == 71
```

If you look in the object's attribute dictionary, you can see that private attributes are actually stored with the names as they appear after the transformation:

```
print(baz.__dict__)

>>>
{'_MyParentObject__private_field': 71}
```

Why doesn't the syntax for private attributes actually enforce strict visibility? The simplest answer is one often-quoted motto of Python: "We are all consenting adults here." What this means is that we don't need the language to prevent us from doing what we want to do. It's our individual choice to extend functionality as we wish and to take responsibility for the consequences of such a risk. Python programmers believe that the benefits of being open—permitting unplanned extension of classes by default—outweigh the downsides.

Beyond that, having the ability to hook language features like attribute access (see Item 47: "Use __getattr__, __getattribute__, and __setattr__ for Lazy Attributes") enables you to mess around with the internals of objects whenever you wish. If you can do that, what is the value of Python trying to prevent private attribute access otherwise?

To minimize damage from accessing internals unknowingly, Python programmers follow a naming convention defined in the style guide (see Item 2: "Follow the PEP 8 Style Guide"). Fields prefixed by a single underscore (like _protected_field) are *protected* by convention, meaning external users of the class should proceed with caution.

However, many programmers who are new to Python use private fields to indicate an internal API that shouldn't be accessed by subclasses or externally:

```
class MyStringClass:
    def __init__(self, value):
        self.__value = value

    def get_value(self):
        return str(self.__value)

foo = MyStringClass(5)
assert foo.get_value() == '5'
```

This is the wrong approach. Inevitably someone—maybe even you—will want to subclass your class to add new behavior or to work around deficiencies in existing methods (e.g., the way that MyStringClass.get_value always returns a string). By choosing private attributes, you're only making subclass overrides and extensions cumbersome and brittle. Your potential subclassers will still access the private fields when they absolutely need to do so:

```
class MyIntegerSubclass(MyStringClass):
    def get_value(self):
        return int(self._MyStringClass__value)
```

```
foo = MyIntegerSubclass('5')
assert foo.get_value() == 5
```

But if the class hierarchy changes beneath you, these classes will break because the private attribute references are no longer valid. Here, the MyIntegerSubclass class's immediate parent, MyStringClass, has had another parent class added, called MyBaseClass:

```
class MyBaseClass:
    def __init__(self, value):
        self.__value = value

    def get_value(self):
        return self.__value

class MyStringClass(MyBaseClass):
    def get_value(self):
        return str(super().get_value())          # Updated

class MyIntegerSubclass(MyStringClass):
    def get_value(self):
        return int(self._MyStringClass__value)  # Not updated
```

The __value attribute is now assigned in the MyBaseClass parent class, not the MyStringClass parent. This causes the private variable reference self._MyStringClass__value to break in MyIntegerSubclass:

```
foo = MyIntegerSubclass(5)
foo.get_value()
```

```
>>>
Traceback ...
AttributeError: 'MyIntegerSubclass' object has no attribute
➥'_MyStringClass__value'
```

In general, it's better to err on the side of allowing subclasses to do more by using protected attributes. Document each protected field and explain which fields are internal APIs available to subclasses and which should be left alone entirely. This is as much advice to other programmers as it is guidance for your future self on how to extend your own code safely:

```
class MyStringClass:
    def __init__(self, value):
        # This stores the user-supplied value for the object.
        # It should be coercible to a string. Once assigned in
        # the object it should be treated as immutable.
```

```
        self._value = value

    ...
```

The only time to seriously consider using private attributes is when you're worried about naming conflicts with subclasses. This problem occurs when a child class unwittingly defines an attribute that was already defined by its parent class:

```
class ApiClass:
    def __init__(self):
        self._value = 5

    def get(self):
        return self._value

class Child(ApiClass):
    def __init__(self):
        super().__init__()
        self._value = 'hello'  # Conflicts

a = Child()
print(f'{a.get()} and {a._value} should be different')

>>>
hello and hello should be different
```

This is primarily a concern with classes that are part of a public API; the subclasses are out of your control, so you can't refactor to fix the problem. Such a conflict is especially possible with attribute names that are very common (like value). To reduce the risk of this issue occurring, you can use a private attribute in the parent class to ensure that there are no attribute names that overlap with child classes:

```
class ApiClass:
    def __init__(self):
        self.__value = 5        # Double underscore

    def get(self):
        return self.__value     # Double underscore

class Child(ApiClass):
    def __init__(self):
        super().__init__()
        self._value = 'hello'  # OK!
```

```
a = Child()
print(f'{a.get()} and {a._value} are different')

>>>
5 and hello are different
```

Things to Remember

✦ Private attributes aren't rigorously enforced by the Python compiler.

✦ Plan from the beginning to allow subclasses to do more with your internal APIs and attributes instead of choosing to lock them out.

✦ Use documentation of protected fields to guide subclasses instead of trying to force access control with private attributes.

✦ Only consider using private attributes to avoid naming conflicts with subclasses that are out of your control.

Item 43: Inherit from collections.abc for Custom Container Types

Much of programming in Python is defining classes that contain data and describing how such objects relate to each other. Every Python class is a container of some kind, encapsulating attributes and functionality together. Python also provides built-in container types for managing data: lists, tuples, sets, and dictionaries.

When you're designing classes for simple use cases like sequences, it's natural to want to subclass Python's built-in list type directly. For example, say I want to create my own custom list type that has additional methods for counting the frequency of its members:

```
class FrequencyList(list):
    def __init__(self, members):
        super().__init__(members)

    def frequency(self):
        counts = {}
        for item in self:
            counts[item] = counts.get(item, 0) + 1
        return counts
```

By subclassing list, I get all of list's standard functionality and preserve the semantics familiar to all Python programmers. I can define additional methods to provide any custom behaviors that I need:

```
foo = FrequencyList(['a', 'b', 'a', 'c', 'b', 'a', 'd'])
print('Length is', len(foo))
```

```
foo.pop()
print('After pop:', repr(foo))
print('Frequency:', foo.frequency())

>>>
Length is 7
After pop: ['a', 'b', 'a', 'c', 'b', 'a']
Frequency: {'a': 3, 'b': 2, 'c': 1}
```

Now, imagine that I want to provide an object that feels like a list and allows indexing but isn't a list subclass. For example, say that I want to provide sequence semantics (like list or tuple) for a binary tree class:

```
class BinaryNode:
    def __init__(self, value, left=None, right=None):
        self.value = value
        self.left = left
        self.right = right
```

How do you make this class act like a sequence type? Python implements its container behaviors with instance methods that have special names. When you access a sequence item by index:

```
bar = [1, 2, 3]
bar[0]
```

it will be interpreted as:

```
bar.__getitem__(0)
```

To make the BinaryNode class act like a sequence, you can provide a custom implementation of __getitem__ (often pronounced "dunder getitem" as an abbreviation for "double underscore getitem") that traverses the object tree depth first:

```
class IndexableNode(BinaryNode):
    def _traverse(self):
        if self.left is not None:
            yield from self.left._traverse()
        yield self
        if self.right is not None:
            yield from self.right._traverse()

    def __getitem__(self, index):
        for i, item in enumerate(self._traverse()):
            if i == index:
                return item.value
        raise IndexError(f'Index {index} is out of range')
```

You can construct your binary tree as usual:

```
tree = IndexableNode(
    10,
    left=IndexableNode(
        5,
        left=IndexableNode(2),
        right=IndexableNode(
            6,
            right=IndexableNode(7))),
    right=IndexableNode(
        15,
        left=IndexableNode(11)))
```

But you can also access it like a list in addition to being able to traverse the tree with the left and right attributes:

```
print('LRR is', tree.left.right.right.value)
print('Index 0 is', tree[0])
print('Index 1 is', tree[1])
print('11 in the tree?', 11 in tree)
print('17 in the tree?', 17 in tree)
print('Tree is', list(tree))
```

```
>>>
LRR is 7
Index 0 is 2
Index 1 is 5
11 in the tree? True
17 in the tree? False
Tree is [2, 5, 6, 7, 10, 11, 15]
```

The problem is that implementing __getitem__ isn't enough to provide all of the sequence semantics you'd expect from a list instance:

```
len(tree)
```

```
>>>
Traceback ...
TypeError: object of type 'IndexableNode' has no len()
```

The len built-in function requires another special method, named __len__, that must have an implementation for a custom sequence type:

```
class SequenceNode(IndexableNode):
    def __len__(self):
        for count, _ in enumerate(self._traverse(), 1):
            pass
        return count
```

```
tree = SequenceNode(
    10,
    left=SequenceNode(
        5,
        left=SequenceNode(2),
        right=SequenceNode(
            6,
            right=SequenceNode(7))),
    right=SequenceNode(
        15,
        left=SequenceNode(11))
)

print('Tree length is', len(tree))

>>>
Tree length is 7
```

Unfortunately, this still isn't enough for the class to fully be a valid sequence. Also missing are the count and index methods that a Python programmer would expect to see on a sequence like list or tuple. It turns out that defining your own container types is much harder than it seems.

To avoid this difficulty throughout the Python universe, the built-in collections.abc module defines a set of abstract base classes that provide all of the typical methods for each container type. When you subclass from these abstract base classes and forget to implement required methods, the module tells you something is wrong:

```
from collections.abc import Sequence

class BadType(Sequence):
    pass

foo = BadType()

>>>
Traceback ...
TypeError: Can't instantiate abstract class BadType with
➥abstract methods __getitem__, __len__
```

When you do implement all the methods required by an abstract base class from collections.abc, as I did above with SequenceNode, it provides all of the additional methods, like index and count, for free:

```
class BetterNode(SequenceNode, Sequence):
    pass
```

```
tree = BetterNode(
    10,
    left=BetterNode(
        5,
        left=BetterNode(2),
        right=BetterNode(
            6,
            right=BetterNode(7))),
    right=BetterNode(
        15,
        left=BetterNode(11))
)

print('Index of 7 is', tree.index(7))
print('Count of 10 is', tree.count(10))

>>>
Index of 7 is 3
Count of 10 is 1
```

The benefit of using these abstract base classes is even greater for more complex container types such as Set and MutableMapping, which have a large number of special methods that need to be implemented to match Python conventions.

Beyond the collections.abc module, Python uses a variety of special methods for object comparisons and sorting, which may be provided by container classes and non-container classes alike (see Item 73: "Know How to Use heapq for Priority Queues" for an example).

Things to Remember

+ Inherit directly from Python's container types (like list or dict) for simple use cases.

+ Beware of the large number of methods required to implement custom container types correctly.

+ Have your custom container types inherit from the interfaces defined in collections.abc to ensure that your classes match required interfaces and behaviors.

Chapter 6

Metaclasses and Attributes

Metaclasses are often mentioned in lists of Python's features, but few understand what they accomplish in practice. The name *meta-class* vaguely implies a concept above and beyond a class. Simply put, metaclasses let you intercept Python's class statement and provide special behavior each time a class is defined.

Similarly mysterious and powerful are Python's built-in features for dynamically customizing attribute accesses. Along with Python's object-oriented constructs, these facilities provide wonderful tools to ease the transition from simple classes to complex ones.

However, with these powers come many pitfalls. Dynamic attributes enable you to override objects and cause unexpected side effects. Metaclasses can create extremely bizarre behaviors that are unapproachable to newcomers. It's important that you follow the *rule of least surprise* and only use these mechanisms to implement well-understood idioms.

Item 44: Use Plain Attributes Instead of Setter and Getter Methods

Programmers coming to Python from other languages may naturally try to implement explicit getter and setter methods in their classes:

```python
class OldResistor:
    def __init__(self, ohms):
        self._ohms = ohms

    def get_ohms(self):
        return self._ohms

    def set_ohms(self, ohms):
        self._ohms = ohms
```

Using these setters and getters is simple, but it's not Pythonic:

```
r0 = OldResistor(50e3)
print('Before:', r0.get_ohms())
r0.set_ohms(10e3)
print('After: ', r0.get_ohms())

>>>
Before: 50000.0
After:   10000.0
```

Such methods are especially clumsy for operations like incrementing in place:

```
r0.set_ohms(r0.get_ohms() - 4e3)
assert r0.get_ohms() == 6e3
```

These utility methods do, however, help define the interface for a class, making it easier to encapsulate functionality, validate usage, and define boundaries. Those are important goals when designing a class to ensure that you don't break callers as the class evolves over time.

In Python, however, you never need to implement explicit setter or getter methods. Instead, you should always start your implementations with simple public attributes, as I do here:

```
class Resistor:
    def __init__(self, ohms):
        self.ohms = ohms
        self.voltage = 0
        self.current = 0

r1 = Resistor(50e3)
r1.ohms = 10e3
```

These attributes make operations like incrementing in place natural and clear:

```
r1.ohms += 5e3
```

Later, if I decide I need special behavior when an attribute is set, I can migrate to the @property decorator (see Item 26: "Define Function Decorators with functools.wraps" for background) and its corresponding setter attribute. Here, I define a new subclass of Resistor that lets me vary the current by assigning the voltage property. Note that in order for this code to work properly, the names of both the setter and the getter methods must match the intended property name:

```
class VoltageResistance(Resistor):
    def __init__(self, ohms):
```

```
        super().__init__(ohms)
        self._voltage = 0

    @property
    def voltage(self):
        return self._voltage

    @voltage.setter
    def voltage(self, voltage):
        self._voltage = voltage
        self.current = self._voltage / self.ohms
```

Now, assigning the voltage property will run the voltage setter method, which in turn will update the current attribute of the object to match:

```
r2 = VoltageResistance(1e3)
print(f'Before: {r2.current:.2f} amps')
r2.voltage = 10
print(f'After:  {r2.current:.2f} amps')
```

```
>>>
Before: 0.00 amps
After:  0.01 amps
```

Specifying a setter on a property also enables me to perform type checking and validation on values passed to the class. Here, I define a class that ensures all resistance values are above zero ohms:

```
class BoundedResistance(Resistor):
    def __init__(self, ohms):
        super().__init__(ohms)

    @property
    def ohms(self):
        return self._ohms

    @ohms.setter
    def ohms(self, ohms):
        if ohms <= 0:
            raise ValueError(f'ohms must be > 0; got {ohms}')
        self._ohms = ohms
```

Assigning an invalid resistance to the attribute now raises an exception:

```
r3 = BoundedResistance(1e3)
r3.ohms = 0
```

```
>>>
Traceback ...
ValueError: ohms must be > 0; got 0
```

An exception is also raised if I pass an invalid value to the constructor:

```
BoundedResistance(-5)
```

```
>>>
Traceback ...
ValueError: ohms must be > 0; got -5
```

This happens because BoundedResistance.__init__ calls Resistor.
__init__, which assigns self.ohms = -5. That assignment causes
the @ohms.setter method from BoundedResistance to be called, and it
immediately runs the validation code before object construction has
completed.

I can even use @property to make attributes from parent classes
immutable:

```
class FixedResistance(Resistor):
    def __init__(self, ohms):
        super().__init__(ohms)

    @property
    def ohms(self):
        return self._ohms

    @ohms.setter
    def ohms(self, ohms):
        if hasattr(self, '_ohms'):
            raise AttributeError("Ohms is immutable")
        self._ohms = ohms
```

Trying to assign to the property after construction raises an exception:

```
r4 = FixedResistance(1e3)
r4.ohms = 2e3
```

```
>>>
Traceback ...
AttributeError: Ohms is immutable
```

When you use @property methods to implement setters and getters,
be sure that the behavior you implement is not surprising. For example, don't set other attributes in getter property methods:

```
class MysteriousResistor(Resistor):
    @property
    def ohms(self):
```

```
        self.voltage = self._ohms * self.current
        return self._ohms

    @ohms.setter
    def ohms(self, ohms):
        self._ohms = ohms
```

Setting other attributes in getter property methods leads to extremely bizarre behavior:

```
r7 = MysteriousResistor(10)
r7.current = 0.01
print(f'Before: {r7.voltage:.2f}')
r7.ohms
print(f'After:  {r7.voltage:.2f}')
```

```
>>>
Before: 0.00
After:  0.10
```

The best policy is to modify only related object state in `@property.setter` methods. Be sure to also avoid any other side effects that the caller may not expect beyond the object, such as importing modules dynamically, running slow helper functions, doing I/O, or making expensive database queries. Users of a class will expect its attributes to be like any other Python object: quick and easy. Use normal methods to do anything more complex or slow.

The biggest shortcoming of `@property` is that the methods for an attribute can only be shared by subclasses. Unrelated classes can't share the same implementation. However, Python also supports *descriptors* (see Item 46: "Use Descriptors for Reusable `@property` Methods") that enable reusable property logic and many other use cases.

Things to Remember

+ Define new class interfaces using simple public attributes and avoid defining setter and getter methods.

+ Use `@property` to define special behavior when attributes are accessed on your objects, if necessary.

+ Follow the rule of least surprise and avoid odd side effects in your `@property` methods.

+ Ensure that `@property` methods are fast; for slow or complex work—especially involving I/O or causing side effects—use normal methods instead.

Item 45: Consider @property Instead of Refactoring Attributes

The built-in @property decorator makes it easy for simple accesses of an instance's attributes to act smarter (see Item 44: "Use Plain Attributes Instead of Setter and Getter Methods"). One advanced but common use of @property is transitioning what was once a simple numerical attribute into an on-the-fly calculation. This is extremely helpful because it lets you migrate all existing usage of a class to have new behaviors without requiring any of the call sites to be rewritten (which is especially important if there's calling code that you don't control). @property also provides an important stopgap for improving interfaces over time.

For example, say that I want to implement a leaky bucket quota using plain Python objects. Here, the Bucket class represents how much quota remains and the duration for which the quota will be available:

```python
from datetime import datetime, timedelta

class Bucket:
    def __init__(self, period):
        self.period_delta = timedelta(seconds=period)
        self.reset_time = datetime.now()
        self.quota = 0

    def __repr__(self):
        return f'Bucket(quota={self.quota})'
```

The leaky bucket algorithm works by ensuring that, whenever the bucket is filled, the amount of quota does not carry over from one period to the next:

```python
def fill(bucket, amount):
    now = datetime.now()
    if (now - bucket.reset_time) > bucket.period_delta:
        bucket.quota = 0
        bucket.reset_time = now
    bucket.quota += amount
```

Each time a quota consumer wants to do something, it must first ensure that it can deduct the amount of quota it needs to use:

```python
def deduct(bucket, amount):
    now = datetime.now()
    if (now - bucket.reset_time) > bucket.period_delta:
        return False  # Bucket hasn't been filled this period
    if bucket.quota - amount < 0:
        return False  # Bucket was filled, but not enough
```

```
    bucket.quota -= amount
    return True       # Bucket had enough, quota consumed
```

To use this class, first I fill the bucket up:

```
bucket = Bucket(60)
fill(bucket, 100)
print(bucket)
```

```
>>>
Bucket(quota=100)
```

Then, I deduct the quota that I need:

```
if deduct(bucket, 99):
    print('Had 99 quota')
else:
    print('Not enough for 99 quota')
print(bucket)
```

```
>>>
Had 99 quota
Bucket(quota=1)
```

Eventually, I'm prevented from making progress because I try to deduct more quota than is available. In this case, the bucket's quota level remains unchanged:

```
if deduct(bucket, 3):
    print('Had 3 quota')
else:
    print('Not enough for 3 quota')
print(bucket)
```

```
>>>
Not enough for 3 quota
Bucket(quota=1)
```

The problem with this implementation is that I never know what quota level the bucket started with. The quota is deducted over the course of the period until it reaches zero. At that point, deduct will always return False until the bucket is refilled. When that happens, it would be useful to know whether callers to deduct are being blocked because the Bucket ran out of quota or because the Bucket never had quota during this period in the first place.

To fix this, I can change the class to keep track of the max_quota issued in the period and the quota_consumed in the period:

```
class NewBucket:
    def __init__(self, period):
        self.period_delta = timedelta(seconds=period)
```

```
        self.reset_time = datetime.now()
        self.max_quota = 0
        self.quota_consumed = 0

    def __repr__(self):
        return (f'NewBucket(max_quota={self.max_quota}, '
                f'quota_consumed={self.quota_consumed})')
```

To match the previous interface of the original Bucket class, I use a @property method to compute the current level of quota on-the-fly using these new attributes:

```
    @property
    def quota(self):
        return self.max_quota - self.quota_consumed
```

When the quota attribute is assigned, I take special action to be compatible with the current usage of the class by the fill and deduct functions:

```
    @quota.setter
    def quota(self, amount):
        delta = self.max_quota - amount
        if amount == 0:
            # Quota being reset for a new period
            self.quota_consumed = 0
            self.max_quota = 0
        elif delta < 0:
            # Quota being filled for the new period
            assert self.quota_consumed == 0
            self.max_quota = amount
        else:
            # Quota being consumed during the period
            assert self.max_quota >= self.quota_consumed
            self.quota_consumed += delta
```

Rerunning the demo code from above produces the same results:

```
bucket = NewBucket(60)
print('Initial', bucket)
fill(bucket, 100)
print('Filled', bucket)

if deduct(bucket, 99):
    print('Had 99 quota')
else:
    print('Not enough for 99 quota')
```

```
print('Now', bucket)

if deduct(bucket, 3):
    print('Had 3 quota')
else:
    print('Not enough for 3 quota')

print('Still', bucket)

>>>
Initial NewBucket(max_quota=0, quota_consumed=0)
Filled NewBucket(max_quota=100, quota_consumed=0)
Had 99 quota
Now NewBucket(max_quota=100, quota_consumed=99)
Not enough for 3 quota
Still NewBucket(max_quota=100, quota_consumed=99)
```

The best part is that the code using Bucket.quota doesn't have to change or know that the class has changed. New usage of Bucket can do the right thing and access max_quota and quota_consumed directly.

I especially like @property because it lets you make incremental progress toward a better data model over time. Reading the Bucket example above, you may have thought that fill and deduct should have been implemented as instance methods in the first place. Although you're probably right (see Item 37: "Compose Classes Instead of Nesting Many Levels of Built-in Types"), in practice there are many situations in which objects start with poorly defined interfaces or act as dumb data containers. This happens when code grows over time, scope increases, multiple authors contribute without anyone considering long-term hygiene, and so on.

@property is a tool to help you address problems you'll come across in real-world code. Don't overuse it. When you find yourself repeatedly extending @property methods, it's probably time to refactor your class instead of further paving over your code's poor design.

Things to Remember

✦ Use @property to give existing instance attributes new functionality.

✦ Make incremental progress toward better data models by using @property.

✦ Consider refactoring a class and all call sites when you find yourself using @property too heavily.

Item 46: Use Descriptors for Reusable @property Methods

The big problem with the @property built-in (see Item 44: "Use Plain Attributes Instead of Setter and Getter Methods" and Item 45: "Consider @property Instead of Refactoring Attributes") is reuse. The methods it decorates can't be reused for multiple attributes of the same class. They also can't be reused by unrelated classes.

For example, say I want a class to validate that the grade received by a student on a homework assignment is a percentage:

```
class Homework:
    def __init__(self):
        self._grade = 0

    @property
    def grade(self):
        return self._grade

    @grade.setter
    def grade(self, value):
        if not (0 <= value <= 100):
            raise ValueError(
                'Grade must be between 0 and 100')
        self._grade = value
```

Using @property makes this class easy to use:

```
galileo = Homework()
galileo.grade = 95
```

Say that I also want to give the student a grade for an exam, where the exam has multiple subjects, each with a separate grade:

```
class Exam:
    def __init__(self):
        self._writing_grade = 0
        self._math_grade = 0

    @staticmethod
    def _check_grade(value):
        if not (0 <= value <= 100):
            raise ValueError(
                'Grade must be between 0 and 100')
```

This quickly gets tedious. For each section of the exam I need to add a new @property and related validation:

```
@property
def writing_grade(self):
    return self._writing_grade

@writing_grade.setter
def writing_grade(self, value):
    self._check_grade(value)
    self._writing_grade = value

@property
def math_grade(self):
    return self._math_grade

@math_grade.setter
def math_grade(self, value):
    self._check_grade(value)
    self._math_grade = value
```

Also, this approach is not general. If I want to reuse this percentage validation in other classes beyond homework and exams, I'll need to write the @property boilerplate and _check_grade method over and over again.

The better way to do this in Python is to use a *descriptor*. The *descriptor protocol* defines how attribute access is interpreted by the language. A descriptor class can provide __get__ and __set__ methods that let you reuse the grade validation behavior without boilerplate. For this purpose, descriptors are also better than mix-ins (see Item 41: "Consider Composing Functionality with Mix-in Classes") because they let you reuse the same logic for many different attributes in a single class.

Here, I define a new class called Exam with class attributes that are Grade instances. The Grade class implements the descriptor protocol:

```
class Grade:
    def __get__(self, instance, instance_type):
        ...

    def __set__(self, instance, value):
        ...
```

```
class Exam:
    # Class attributes
    math_grade = Grade()
    writing_grade = Grade()
    science_grade = Grade()
```

Before I explain how the Grade class works, it's important to understand what Python will do when such descriptor attributes are accessed on an Exam instance. When I assign a property:

```
exam = Exam()
exam.writing_grade = 40
```

it is interpreted as:

```
Exam.__dict__['writing_grade'].__set__(exam, 40)
```

When I retrieve a property:

```
exam.writing_grade
```

it is interpreted as:

```
Exam.__dict__['writing_grade'].__get__(exam, Exam)
```

What drives this behavior is the __getattribute__ method of object (see Item 47: "Use __getattr__, __getattribute__, and __setattr__ for Lazy Attributes"). In short, when an Exam instance doesn't have an attribute named writing_grade, Python falls back to the Exam class's attribute instead. If this class attribute is an object that has __get__ and __set__ methods, Python assumes that you want to follow the descriptor protocol.

Knowing this behavior and how I used @property for grade validation in the Homework class, here's a reasonable first attempt at implementing the Grade descriptor:

```
class Grade:
    def __init__(self):
        self._value = 0

    def __get__(self, instance, instance_type):
        return self._value

    def __set__(self, instance, value):
        if not (0 <= value <= 100):
            raise ValueError(
                'Grade must be between 0 and 100')
        self._value = value
```

Unfortunately, this is wrong and results in broken behavior. Accessing multiple attributes on a single Exam instance works as expected:

```
class Exam:
    math_grade = Grade()
    writing_grade = Grade()
    science_grade = Grade()

first_exam = Exam()
first_exam.writing_grade = 82
first_exam.science_grade = 99
print('Writing', first_exam.writing_grade)
print('Science', first_exam.science_grade)

>>>
Writing 82
Science 99
```

But accessing these attributes on multiple Exam instances causes unexpected behavior:

```
second_exam = Exam()
second_exam.writing_grade = 75
print(f'Second {second_exam.writing_grade} is right')
print(f'First  {first_exam.writing_grade} is wrong; '
      f'should be 82')

>>>
Second 75 is right
First  75 is wrong; should be 82
```

The problem is that a single Grade instance is shared across all Exam instances for the class attribute writing_grade. The Grade instance for this attribute is constructed once in the program lifetime, when the Exam class is first defined, not each time an Exam instance is created.

To solve this, I need the Grade class to keep track of its value for each unique Exam instance. I can do this by saving the per-instance state in a dictionary:

```
class Grade:
    def __init__(self):
        self._values = {}

    def __get__(self, instance, instance_type):
        if instance is None:
            return self
        return self._values.get(instance, 0)
```

```
def __set__(self, instance, value):
    if not (0 <= value <= 100):
        raise ValueError(
            'Grade must be between 0 and 100')
    self._values[instance] = value
```

This implementation is simple and works well, but there's still one gotcha: It leaks memory. The _values dictionary holds a reference to every instance of Exam ever passed to __set__ over the lifetime of the program. This causes instances to never have their reference count go to zero, preventing cleanup by the garbage collector (see Item 81: "Use tracemalloc to Understand Memory Usage and Leaks" for how to detect this type of problem).

To fix this, I can use Python's weakref built-in module. This module provides a special class called WeakKeyDictionary that can take the place of the simple dictionary used for _values. The unique behavior of WeakKeyDictionary is that it removes Exam instances from its set of items when the Python runtime knows it's holding the instance's last remaining reference in the program. Python does the bookkeeping for me and ensures that the _values dictionary will be empty when all Exam instances are no longer in use:

```
from weakref import WeakKeyDictionary

class Grade:
    def __init__(self):
        self._values = WeakKeyDictionary()

    def __get__(self, instance, instance_type):
        ...

    def __set__(self, instance, value):
        ...
```

Using this implementation of the Grade descriptor, everything works as expected:

```
class Exam:
    math_grade = Grade()
    writing_grade = Grade()
    science_grade = Grade()

first_exam = Exam()
first_exam.writing_grade = 82
second_exam = Exam()
second_exam.writing_grade = 75
print(f'First  {first_exam.writing_grade} is right')
print(f'Second {second_exam.writing_grade} is right')
```

```
>>>
First  82 is right
Second 75 is right
```

Things to Remember

✦ Reuse the behavior and validation of @property methods by defining your own descriptor classes.

✦ Use WeakKeyDictionary to ensure that your descriptor classes don't cause memory leaks.

✦ Don't get bogged down trying to understand exactly how __getattribute__ uses the descriptor protocol for getting and setting attributes.

Item 47: Use __getattr__, __getattribute__, and __setattr__ for Lazy Attributes

Python's object hooks make it easy to write generic code for gluing systems together. For example, say that I want to represent the records in a database as Python objects. The database has its schema set already. My code that uses objects corresponding to those records must also know what the database looks like. However, in Python, the code that connects Python objects to the database doesn't need to explicitly specify the schema of the records; it can be generic.

How is that possible? Plain instance attributes, @property methods, and descriptors can't do this because they all need to be defined in advance. Python makes this dynamic behavior possible with the __getattr__ special method. If a class defines __getattr__, that method is called every time an attribute can't be found in an object's instance dictionary:

```
class LazyRecord:
    def __init__(self):
        self.exists = 5

    def __getattr__(self, name):
        value = f'Value for {name}'
        setattr(self, name, value)
        return value
```

Here, I access the missing property foo. This causes Python to call the __getattr__ method above, which mutates the instance dictionary __dict__:

```
data = LazyRecord()
print('Before:', data.__dict__)
```

```
print('foo:    ', data.foo)
print('After: ', data.__dict__)
```

```
>>>
Before: {'exists': 5}
foo:    Value for foo
After: {'exists': 5, 'foo': 'Value for foo'}
```

Here, I add logging to LazyRecord to show when __getattr__ is actually called. Note how I call super().__getattr__() to use the superclass's implementation of __getattr__ in order to fetch the real property value and avoid infinite recursion (see Item 40: "Initialize Parent Classes with super" for background):

```
class LoggingLazyRecord(LazyRecord):
    def __getattr__(self, name):
        print(f'* Called __getattr__({name!r}), '
              f'populating instance dictionary')
        result = super().__getattr__(name)
        print(f'* Returning {result!r}')
        return result
```

```
data = LoggingLazyRecord()
print('exists:     ', data.exists)
print('First foo:  ', data.foo)
print('Second foo: ', data.foo)
```

```
>>>
exists:      5
* Called __getattr__('foo'), populating instance dictionary
* Returning 'Value for foo'
First foo:   Value for foo
Second foo:  Value for foo
```

The exists attribute is present in the instance dictionary, so __getattr__ is never called for it. The foo attribute is not in the instance dictionary initially, so __getattr__ is called the first time. But the call to __getattr__ for foo also does a setattr, which populates foo in the instance dictionary. This is why the second time I access foo, it doesn't log a call to __getattr__.

This behavior is especially helpful for use cases like lazily accessing schemaless data. __getattr__ runs once to do the hard work of loading a property; all subsequent accesses retrieve the existing result.

Say that I also want transactions in this database system. The next time the user accesses a property, I want to know whether the corresponding record in the database is still valid and whether the transaction is still open. The __getattr__ hook won't let me do this reliably because it will use the object's instance dictionary as the fast path for existing attributes.

To enable this more advanced use case, Python has another object hook called __getattribute__. This special method is called every time an attribute is accessed on an object, even in cases where it *does* exist in the attribute dictionary. This enables me to do things like check global transaction state on every property access. It's important to note that such an operation can incur significant overhead and negatively impact performance, but sometimes it's worth it. Here, I define ValidatingRecord to log each time __getattribute__ is called:

```python
class ValidatingRecord:
    def __init__(self):
        self.exists = 5

    def __getattribute__(self, name):
        print(f'* Called __getattribute__({name!r})')
        try:
            value = super().__getattribute__(name)
            print(f'* Found {name!r}, returning {value!r}')
            return value
        except AttributeError:
            value = f'Value for {name}'
            print(f'* Setting {name!r} to {value!r}')
            setattr(self, name, value)
            return value


data = ValidatingRecord()
print('exists:     ', data.exists)
print('First foo: ', data.foo)
print('Second foo: ', data.foo)

>>>
* Called __getattribute__('exists')
* Found 'exists', returning 5
exists:      5
* Called __getattribute__('foo')
* Setting 'foo' to 'Value for foo'
First foo:   Value for foo
* Called __getattribute__('foo')
* Found 'foo', returning 'Value for foo'
Second foo:  Value for foo
```

In the event that a dynamically accessed property shouldn't exist, I can raise an AttributeError to cause Python's standard missing property behavior for both __getattr__ and __getattribute__:

```
class MissingPropertyRecord:
    def __getattr__(self, name):
        if name == 'bad_name':
            raise AttributeError(f'{name} is missing')
        ...

data = MissingPropertyRecord()
data.bad_name

>>>
Traceback ...
AttributeError: bad_name is missing
```

Python code implementing generic functionality often relies on the hasattr built-in function to determine when properties exist, and the getattr built-in function to retrieve property values. These functions also look in the instance dictionary for an attribute name before calling __getattr__:

```
data = LoggingLazyRecord()  # Implements __getattr__
print('Before:          ', data.__dict__)
print('Has first foo:  ', hasattr(data, 'foo'))
print('After:           ', data.__dict__)
print('Has second foo: ', hasattr(data, 'foo'))

>>>
Before:         {'exists': 5}
* Called __getattr__('foo'), populating instance dictionary
* Returning 'Value for foo'
Has first foo:   True
After:          {'exists': 5, 'foo': 'Value for foo'}
Has second foo:  True
```

In the example above, __getattr__ is called only once. In contrast, classes that implement __getattribute__ have that method called each time hasattr or getattr is used with an instance:

```
data = ValidatingRecord()  # Implements __getattribute__
print('Has first foo:  ', hasattr(data, 'foo'))
print('Has second foo: ', hasattr(data, 'foo'))

>>>
* Called __getattribute__('foo')
* Setting 'foo' to 'Value for foo'
Has first foo:   True
```

```
* Called __getattribute__('foo')
* Found 'foo', returning 'Value for foo'
Has second foo:  True
```

Now, say that I want to lazily push data back to the database when values are assigned to my Python object. I can do this with __setattr__, a similar object hook that lets you intercept arbitrary attribute assignments. Unlike when retrieving an attribute with __getattr__ and __getattribute__, there's no need for two separate methods. The __setattr__ method is always called every time an attribute is assigned on an instance (either directly or through the setattr built-in function):

```
class SavingRecord:
    def __setattr__(self, name, value):
        # Save some data for the record
        ...
        super().__setattr__(name, value)
```

Here, I define a logging subclass of SavingRecord. Its __setattr__ method is always called on each attribute assignment:

```
class LoggingSavingRecord(SavingRecord):
    def __setattr__(self, name, value):
        print(f'* Called __setattr__({name!r}, {value!r})')
        super().__setattr__(name, value)


data = LoggingSavingRecord()
print('Before: ', data.__dict__)
data.foo = 5
print('After:  ', data.__dict__)
data.foo = 7
print('Finally:', data.__dict__)

>>>
Before:  {}
* Called __setattr__('foo', 5)
After:   {'foo': 5}
* Called __setattr__('foo', 7)
Finally: {'foo': 7}
```

The problem with __getattribute__ and __setattr__ is that they're called on every attribute access for an object, even when you may not want that to happen. For example, say that I want attribute accesses on my object to actually look up keys in an associated dictionary:

```
class BrokenDictionaryRecord:
    def __init__(self, data):
        self._data = {}
```

```
    def __getattribute__(self, name):
        print(f'* Called __getattribute__({name!r})')
        return self._data[name]
```

This requires accessing self._data from the __getattribute__ method. However, if I actually try to do that, Python will recurse until it reaches its stack limit, and then it'll die:

```
data = BrokenDictionaryRecord({'foo': 3})
data.foo
```

```
>>>
* Called __getattribute__('foo')
* Called __getattribute__('_data')
* Called __getattribute__('_data')
* Called __getattribute__('_data')
...
Traceback ...
RecursionError: maximum recursion depth exceeded while calling
➥a Python object
```

The problem is that __getattribute__ accesses self._data, which causes __getattribute__ to run again, which accesses self._data again, and so on. The solution is to use the super().__getattribute__ method to fetch values from the instance attribute dictionary. This avoids the recursion:

```
class DictionaryRecord:
    def __init__(self, data):
        self._data = data

    def __getattribute__(self, name):
        print(f'* Called __getattribute__({name!r})')
        data_dict = super().__getattribute__('_data')
        return data_dict[name]
```

```
data = DictionaryRecord({'foo': 3})
print('foo: ', data.foo)
```

```
>>>
* Called __getattribute__('foo')
foo:  3
```

__setattr__ methods that modify attributes on an object also need to use super().__setattr__ accordingly.

Things to Remember

✦ Use __getattr__ and __setattr__ to lazily load and save attributes for an object.

+ Understand that __getattr__ only gets called when accessing a missing attribute, whereas __getattribute__ gets called every time any attribute is accessed.

+ Avoid infinite recursion in __getattribute__ and __setattr__ by using methods from super() (i.e., the object class) to access instance attributes.

Item 48: Validate Subclasses with __init_subclass__

One of the simplest applications of metaclasses is verifying that a class was defined correctly. When you're building a complex class hierarchy, you may want to enforce style, require overriding methods, or have strict relationships between class attributes. Metaclasses enable these use cases by providing a reliable way to run your validation code each time a new subclass is defined.

Often a class's validation code runs in the __init__ method, when an object of the class's type is constructed at runtime (see Item 44: "Use Plain Attributes Instead of Setter and Getter Methods" for an example). Using metaclasses for validation can raise errors much earlier, such as when the module containing the class is first imported at program startup.

Before I get into how to define a metaclass for validating subclasses, it's important to understand the metaclass action for standard objects. A metaclass is defined by inheriting from type. In the default case, a metaclass receives the contents of associated class statements in its __new__ method. Here, I can inspect and modify the class information before the type is actually constructed:

```python
class Meta(type):
    def __new__(meta, name, bases, class_dict):
        print(f'* Running {meta}.__new__ for {name}')
        print('Bases:', bases)
        print(class_dict)
        return type.__new__(meta, name, bases, class_dict)

class MyClass(metaclass=Meta):
    stuff = 123

    def foo(self):
        pass

class MySubclass(MyClass):
    other = 567

    def bar(self):
        pass
```

The metaclass has access to the name of the class, the parent classes it inherits from (bases), and all the class attributes that were defined in the class's body. All classes inherit from object, so it's not explicitly listed in the tuple of base classes:

```
>>>
* Running <class '__main__.Meta'>.__new__ for MyClass
Bases: ()
{'__module__': '__main__',
 '__qualname__': 'MyClass',
 'stuff': 123,
 'foo': <function MyClass.foo at 0x105a05280>}
* Running <class '__main__.Meta'>.__new__ for MySubclass
Bases: (<class '__main__.MyClass'>,)
{'__module__': '__main__',
 '__qualname__': 'MySubclass',
 'other': 567,
 'bar': <function MySubclass.bar at 0x105a05310>}
```

I can add functionality to the Meta.__new__ method in order to validate all of the parameters of an associated class before it's defined. For example, say that I want to represent any type of multisided polygon. I can do this by defining a special validating metaclass and using it in the base class of my polygon class hierarchy. Note that it's important not to apply the same validation to the base class:

```
class ValidatePolygon(type):
    def __new__(meta, name, bases, class_dict):
        # Only validate subclasses of the Polygon class
        if bases:
            if class_dict['sides'] < 3:
                raise ValueError('Polygons need 3+ sides')
        return type.__new__(meta, name, bases, class_dict)

class Polygon(metaclass=ValidatePolygon):
    sides = None  # Must be specified by subclasses

    @classmethod
    def interior_angles(cls):
        return (cls.sides - 2) * 180

class Triangle(Polygon):
    sides = 3
```

```
class Rectangle(Polygon):
    sides = 4

class Nonagon(Polygon):
    sides = 9

assert Triangle.interior_angles() == 180
assert Rectangle.interior_angles() == 360
assert Nonagon.interior_angles() == 1260
```

If I try to define a polygon with fewer than three sides, the validation will cause the class statement to fail immediately after the class statement body. This means the program will not even be able to start running when I define such a class (unless it's defined in a dynamically imported module; see Item 88: "Know How to Break Circular Dependencies" for how this can happen):

```
print('Before class')

class Line(Polygon):
    print('Before sides')
    sides = 2
    print('After sides')

print('After class')

>>>
Before class
Before sides
After sides
Traceback ...
ValueError: Polygons need 3+ sides
```

This seems like quite a lot of machinery in order to get Python to accomplish such a basic task. Luckily, Python 3.6 introduced simplified syntax—the __init_subclass__ special class method—for achieving the same behavior while avoiding metaclasses entirely. Here, I use this mechanism to provide the same level of validation as before:

```
class BetterPolygon:
    sides = None  # Must be specified by subclasses

    def __init_subclass__(cls):
        super().__init_subclass__()
        if cls.sides < 3:
            raise ValueError('Polygons need 3+ sides')
```

```
    @classmethod
    def interior_angles(cls):
        return (cls.sides - 2) * 180

class Hexagon(BetterPolygon):
    sides = 6

assert Hexagon.interior_angles() == 720
```

The code is much shorter now, and the ValidatePolygon metaclass is gone entirely. It's also easier to follow since I can access the sides attribute directly on the cls instance in __init_subclass__ instead of having to go into the class's dictionary with class_dict['sides']. If I define an invalid subclass of BetterPolygon, the same exception is raised:

```
print('Before class')

class Point(BetterPolygon):
    sides = 1

print('After class')

>>>
Before class
Traceback ...
ValueError: Polygons need 3+ sides
```

Another problem with the standard Python metaclass machinery is that you can only specify a single metaclass per class definition. Here, I define a second metaclass that I'd like to use for validating the fill color used for a region (not necessarily just polygons):

```
class ValidateFilled(type):
    def __new__(meta, name, bases, class_dict):
        # Only validate subclasses of the Filled class
        if bases:
            if class_dict['color'] not in ('red', 'green'):
                raise ValueError('Fill color must be supported')
        return type.__new__(meta, name, bases, class_dict)

class Filled(metaclass=ValidateFilled):
    color = None  # Must be specified by subclasses
```

When I try to use the Polygon metaclass and Filled metaclass together, I get a cryptic error message:

```
class RedPentagon(Filled, Polygon):
    color = 'red'
    sides = 5
```

```
>>>
Traceback ...
TypeError: metaclass conflict: the metaclass of a derived
➥class must be a (non-strict) subclass of the metaclasses
➥of all its bases
```

It's possible to fix this by creating a complex hierarchy of metaclass type definitions to layer validation:

```
class ValidatePolygon(type):
    def __new__(meta, name, bases, class_dict):
        # Only validate non-root classes
        if not class_dict.get('is_root'):
            if class_dict['sides'] < 3:
                raise ValueError('Polygons need 3+ sides')
        return type.__new__(meta, name, bases, class_dict)

class Polygon(metaclass=ValidatePolygon):
    is_root = True
    sides = None  # Must be specified by subclasses

class ValidateFilledPolygon(ValidatePolygon):
    def __new__(meta, name, bases, class_dict):
        # Only validate non-root classes
        if not class_dict.get('is_root'):
            if class_dict['color'] not in ('red', 'green'):
                raise ValueError('Fill color must be supported')
        return super().__new__(meta, name, bases, class_dict)

class FilledPolygon(Polygon, metaclass=ValidateFilledPolygon):
    is_root = True
    color = None  # Must be specified by subclasses
```

This requires every FilledPolygon instance to be a Polygon instance:

```
class GreenPentagon(FilledPolygon):
    color = 'green'
    sides = 5

greenie = GreenPentagon()
assert isinstance(greenie, Polygon)
```

Validation works for colors:

```
class OrangePentagon(FilledPolygon):
    color = 'orange'
    sides = 5
```

```
>>>
Traceback ...
ValueError: Fill color must be supported
```

Validation also works for number of sides:

```
class RedLine(FilledPolygon):
    color = 'red'
    sides = 2
```

```
>>>
Traceback ...
ValueError: Polygons need 3+ sides
```

However, this approach ruins composability, which is often the purpose of class validation like this (similar to mix-ins; see Item 41: "Consider Composing Functionality with Mix-in Classes"). If I want to apply the color validation logic from ValidateFilledPolygon to another hierarchy of classes, I'll have to duplicate all of the logic again, which reduces code reuse and increases boilerplate.

The __init_subclass__ special class method can also be used to solve this problem. It can be defined by multiple levels of a class hierarchy as long as the super built-in function is used to call any parent or sibling __init_subclass__ definitions (see Item 40: "Initialize Parent Classes with super" for a similar example). It's even compatible with multiple inheritance. Here, I define a class to represent region fill color that can be composed with the BetterPolygon class from before:

```
class Filled:
    color = None  # Must be specified by subclasses

    def __init_subclass__(cls):
        super().__init_subclass__()
        if cls.color not in ('red', 'green', 'blue'):
            raise ValueError('Fills need a valid color')
```

I can inherit from both classes to define a new class. Both classes call super().__init_subclass__(), causing their corresponding validation logic to run when the subclass is created:

```
class RedTriangle(Filled, Polygon):
    color = 'red'
    sides = 3
```

```
ruddy = RedTriangle()
assert isinstance(ruddy, Filled)
assert isinstance(ruddy, Polygon)
```

If I specify the number of sides incorrectly, I get a validation error:

```
print('Before class')

class BlueLine(Filled, Polygon):
    color = 'blue'
    sides = 2

print('After class')
```

```
>>>
Before class
Traceback ...
ValueError: Polygons need 3+ sides
```

If I specify the color incorrectly, I also get a validation error:

```
print('Before class')

class BeigeSquare(Filled, Polygon):
    color = 'beige'
    sides = 4

print('After class')
```

```
>>>
Before class
Traceback ...
ValueError: Fills need a valid color
```

You can even use __init_subclass__ in complex cases like diamond inheritance (see Item 40: "Initialize Parent Classes with super"). Here, I define a basic diamond hierarchy to show this in action:

```
class Top:
    def __init_subclass__(cls):
        super().__init_subclass__()
        print(f'Top for {cls}')

class Left(Top):
    def __init_subclass__(cls):
        super().__init_subclass__()
        print(f'Left for {cls}')

class Right(Top):
    def __init_subclass__(cls):
```

```
        super().__init_subclass__()
        print(f'Right for {cls}')

class Bottom(Left, Right):
    def __init_subclass__(cls):
        super().__init_subclass__()
        print(f'Bottom for {cls}')

>>>
Top for <class '__main__.Left'>
Top for <class '__main__.Right'>
Top for <class '__main__.Bottom'>
Right for <class '__main__.Bottom'>
Left for <class '__main__.Bottom'>
```

As expected, Top.__init_subclass__ is called only a single time for each class, even though there are two paths to it for the Bottom class through its Left and Right parent classes.

Things to Remember

✦ The __new__ method of metaclasses is run after the class statement's entire body has been processed.

✦ Metaclasses can be used to inspect or modify a class after it's defined but before it's created, but they're often more heavyweight than what you need.

✦ Use __init_subclass__ to ensure that subclasses are well formed at the time they are defined, before objects of their type are constructed.

✦ Be sure to call super().__init_subclass__ from within your class's __init_subclass__ definition to enable validation in multiple layers of classes and multiple inheritance.

Item 49: Register Class Existence with __init_subclass__

Another common use of metaclasses is to automatically register types in a program. Registration is useful for doing reverse lookups, where you need to map a simple identifier back to a corresponding class.

For example, say that I want to implement my own serialized representation of a Python object using JSON. I need a way to turn an object into a JSON string. Here, I do this generically by defining a base class that records the constructor parameters and turns them into a JSON dictionary:

```
import json

class Serializable:
    def __init__(self, *args):
        self.args = args

    def serialize(self):
        return json.dumps({'args': self.args})
```

This class makes it easy to serialize simple, immutable data structures like Point2D to a string:

```
class Point2D(Serializable):
    def __init__(self, x, y):
        super().__init__(x, y)
        self.x = x
        self.y = y

    def __repr__(self):
        return f'Point2D({self.x}, {self.y})'

point = Point2D(5, 3)
print('Object:   ', point)
print('Serialized:', point.serialize())

>>>
Object:    Point2D(5, 3)
Serialized: {"args": [5, 3]}
```

Now, I need to deserialize this JSON string and construct the Point2D object it represents. Here, I define another class that can deserialize the data from its Serializable parent class:

```
class Deserializable(Serializable):
    @classmethod
    def deserialize(cls, json_data):
        params = json.loads(json_data)
        return cls(*params['args'])
```

Using Deserializable makes it easy to serialize and deserialize simple, immutable objects in a generic way:

```
class BetterPoint2D(Deserializable):
    ...
```

```
before = BetterPoint2D(5, 3)
print('Before:    ', before)
data = before.serialize()
print('Serialized:', data)
after = BetterPoint2D.deserialize(data)
print('After:     ', after)

>>>
Before:     Point2D(5, 3)
Serialized: {"args": [5, 3]}
After:      Point2D(5, 3)
```

The problem with this approach is that it works only if you know the intended type of the serialized data ahead of time (e.g., Point2D, BetterPoint2D). Ideally, you'd have a large number of classes serializing to JSON and one common function that could deserialize any of them back to a corresponding Python object.

To do this, I can include the serialized object's class name in the JSON data:

```
class BetterSerializable:
    def __init__(self, *args):
        self.args = args

    def serialize(self):
        return json.dumps({
            'class': self.__class__.__name__,
            'args': self.args,
        })

    def __repr__(self):
        name = self.__class__.__name__
        args_str = ', '.join(str(x) for x in self.args)
        return f'{name}({args_str})'
```

Then, I can maintain a mapping of class names back to constructors for those objects. The general deserialize function works for any classes passed to register_class:

```
registry = {}

def register_class(target_class):
    registry[target_class.__name__] = target_class

def deserialize(data):
    params = json.loads(data)
```

```
name = params['class']
target_class = registry[name]
return target_class(*params['args'])
```

To ensure that deserialize always works properly, I must call register_class for every class I may want to deserialize in the future:

```
class EvenBetterPoint2D(BetterSerializable):
    def __init__(self, x, y):
        super().__init__(x, y)
        self.x = x
        self.y = y
```

```
register_class(EvenBetterPoint2D)
```

Now, I can deserialize an arbitrary JSON string without having to know which class it contains:

```
before = EvenBetterPoint2D(5, 3)
print('Before:    ', before)
data = before.serialize()
print('Serialized:', data)
after = deserialize(data)
print('After:     ', after)
```

```
>>>
Before:     EvenBetterPoint2D(5, 3)
Serialized: {"class": "EvenBetterPoint2D", "args": [5, 3]}
After:      EvenBetterPoint2D(5, 3)
```

The problem with this approach is that it's possible to forget to call register_class:

```
class Point3D(BetterSerializable):
    def __init__(self, x, y, z):
        super().__init__(x, y, z)
        self.x = x
        self.y = y
        self.z = z
```

```
# Forgot to call register_class! Whoops!
```

This causes the code to break at runtime, when I finally try to deserialize an instance of a class I forgot to register:

```
point = Point3D(5, 9, -4)
data = point.serialize()
deserialize(data)
```

```
>>>
Traceback ...
KeyError: 'Point3D'
```

Even though I chose to subclass BetterSerializable, I don't actually get all of its features if I forget to call register_class after the class statement body. This approach is error prone and especially challenging for beginners. The same omission can happen with *class decorators* (see Item 51: "Prefer Class Decorators Over Metaclasses for Composable Class Extensions" for when those are appropriate).

What if I could somehow act on the programmer's intent to use BetterSerializable and ensure that register_class is called in all cases? Metaclasses enable this by intercepting the class statement when subclasses are defined (see Item 48: "Validate Subclasses with __init_subclass__" for details on the machinery). Here, I use a metaclass to register the new type immediately after the class's body:

```
class Meta(type):
    def __new__(meta, name, bases, class_dict):
        cls = type.__new__(meta, name, bases, class_dict)
        register_class(cls)
        return cls


class RegisteredSerializable(BetterSerializable,
                            metaclass=Meta):
    pass
```

When I define a subclass of RegisteredSerializable, I can be confident that the call to register_class happened and deserialize will always work as expected:

```
class Vector3D(RegisteredSerializable):
    def __init__(self, x, y, z):
        super().__init__(x, y, z)
        self.x, self.y, self.z = x, y, z


before = Vector3D(10, -7, 3)
print('Before:   ', before)
data = before.serialize()
print('Serialized:', data)
print('After:    ', deserialize(data))
```

```
>>>
Before:    Vector3D(10, -7, 3)
Serialized: {"class": "Vector3D", "args": [10, -7, 3]}
After:     Vector3D(10, -7, 3)
```

An even better approach is to use the __init_subclass__ special class method. This simplified syntax, introduced in Python 3.6, reduces the visual noise of applying custom logic when a class is defined. It also makes it more approachable to beginners who may be confused by the complexity of metaclass syntax:

```
class BetterRegisteredSerializable(BetterSerializable):
    def __init_subclass__(cls):
        super().__init_subclass__()
        register_class(cls)

class Vector1D(BetterRegisteredSerializable):
    def __init__(self, magnitude):
        super().__init__(magnitude)
        self.magnitude = magnitude

before = Vector1D(6)
print('Before:    ', before)
data = before.serialize()
print('Serialized:', data)
print('After:     ', deserialize(data))

>>>
Before:     Vector1D(6)
Serialized: {"class": "Vector1D", "args": [6]}
After:      Vector1D(6)
```

By using __init_subclass__ (or metaclasses) for class registration, you can ensure that you'll never miss registering a class as long as the inheritance tree is right. This works well for serialization, as I've shown, and also applies to database object-relational mappings (ORMs), extensible plug-in systems, and callback hooks.

Things to Remember

✦ Class registration is a helpful pattern for building modular Python programs.

✦ Metaclasses let you run registration code automatically each time a base class is subclassed in a program.

✦ Using metaclasses for class registration helps you avoid errors by ensuring that you never miss a registration call.

✦ Prefer __init_subclass__ over standard metaclass machinery because it's clearer and easier for beginners to understand.

Item 50: Annotate Class Attributes with __set_name__

One more useful feature enabled by metaclasses is the ability to modify or annotate properties after a class is defined but before the class is actually used. This approach is commonly used with *descriptors* (see Item 46: "Use Descriptors for Reusable @property Methods") to give them more introspection into how they're being used within their containing class.

For example, say that I want to define a new class that represents a row in a customer database. I'd like to have a corresponding property on the class for each column in the database table. Here, I define a descriptor class to connect attributes to column names:

```
class Field:
    def __init__(self, name):
        self.name = name
        self.internal_name = '_' + self.name

    def __get__(self, instance, instance_type):
        if instance is None:
            return self
        return getattr(instance, self.internal_name, '')

    def __set__(self, instance, value):
        setattr(instance, self.internal_name, value)
```

With the column name stored in the Field descriptor, I can save all of the per-instance state directly in the instance dictionary as protected fields by using the setattr built-in function, and later I can load state with getattr. At first, this seems to be much more convenient than building descriptors with the weakref built-in module to avoid memory leaks.

Defining the class representing a row requires supplying the database table's column name for each class attribute:

```
class Customer:
    # Class attributes
    first_name = Field('first_name')
    last_name = Field('last_name')
    prefix = Field('prefix')
    suffix = Field('suffix')
```

Using the class is simple. Here, you can see how the Field descriptors modify the instance dictionary __dict__ as expected:

```
cust = Customer()
print(f'Before: {cust.first_name!r} {cust.__dict__}')
```

```
cust.first_name = 'Euclid'
print(f'After:  {cust.first_name!r} {cust.__dict__}')
```

```
>>>
Before: '' {}
After:  'Euclid' {'_first_name': 'Euclid'}
```

But the class definition seems redundant. I already declared the name of the field for the class on the left ('field_name ='). Why do I also have to pass a string containing the same information to the Field constructor (Field('first_name')) on the right?

```
class Customer:
    # Left side is redundant with right side
    first_name = Field('first_name')
    ...
```

The problem is that the order of operations in the Customer class definition is the opposite of how it reads from left to right. First, the Field constructor is called as Field('first_name'). Then, the return value of that is assigned to Customer.field_name. There's no way for a Field instance to know upfront which class attribute it will be assigned to.

To eliminate this redundancy, I can use a metaclass. Metaclasses let you hook the class statement directly and take action as soon as a class body is finished (see Item 48: "Validate Subclasses with __init_subclass__" for details on how they work). In this case, I can use the metaclass to assign Field.name and Field.internal_name on the descriptor automatically instead of manually specifying the field name multiple times:

```
class Meta(type):
    def __new__(meta, name, bases, class_dict):
        for key, value in class_dict.items():
            if isinstance(value, Field):
                value.name = key
                value.internal_name = '_' + key
        cls = type.__new__(meta, name, bases, class_dict)
        return cls
```

Here, I define a base class that uses the metaclass. All classes representing database rows should inherit from this class to ensure that they use the metaclass:

```
class DatabaseRow(metaclass=Meta):
    pass
```

To work with the metaclass, the Field descriptor is largely unchanged. The only difference is that it no longer requires arguments to be passed to its constructor. Instead, its attributes are set by the Meta.__new__ method above:

```
class Field:
    def __init__(self):
        # These will be assigned by the metaclass.
        self.name = None
        self.internal_name = None

    def __get__(self, instance, instance_type):
        if instance is None:
            return self
        return getattr(instance, self.internal_name, '')

    def __set__(self, instance, value):
        setattr(instance, self.internal_name, value)
```

By using the metaclass, the new DatabaseRow base class, and the new Field descriptor, the class definition for a database row no longer has the redundancy from before:

```
class BetterCustomer(DatabaseRow):
    first_name = Field()
    last_name = Field()
    prefix = Field()
    suffix = Field()
```

The behavior of the new class is identical to the behavior of the old one:

```
cust = BetterCustomer()
print(f'Before: {cust.first_name!r} {cust.__dict__}')
cust.first_name = 'Euler'
print(f'After:  {cust.first_name!r} {cust.__dict__}')

>>>
Before: '' {}
After:  'Euler' {'_first_name': 'Euler'}
```

The trouble with this approach is that you can't use the Field class for properties unless you also inherit from DatabaseRow. If you somehow forget to subclass DatabaseRow, or if you don't want to due to other structural requirements of the class hierarchy, the code will break:

```
class BrokenCustomer:
    first_name = Field()
    last_name = Field()
    prefix = Field()
    suffix = Field()
```

```
cust = BrokenCustomer()
cust.first_name = 'Mersenne'
```

```
>>>
Traceback ...
TypeError: attribute name must be string, not 'NoneType'
```

The solution to this problem is to use the __set_name__ special method for descriptors. This method, introduced in Python 3.6, is called on every descriptor instance when its containing class is defined. It receives as parameters the owning class that contains the descriptor instance and the attribute name to which the descriptor instance was assigned. Here, I avoid defining a metaclass entirely and move what the Meta.__new__ method from above was doing into __set_name__:

```
class Field:
    def __init__(self):
        self.name = None
        self.internal_name = None

    def __set_name__(self, owner, name):
        # Called on class creation for each descriptor
        self.name = name
        self.internal_name = '_' + name

    def __get__(self, instance, instance_type):
        if instance is None:
            return self
        return getattr(instance, self.internal_name, '')

    def __set__(self, instance, value):
        setattr(instance, self.internal_name, value)
```

Now, I can get the benefits of the Field descriptor without having to inherit from a specific parent class or having to use a metaclass:

```
class FixedCustomer:
    first_name = Field()
    last_name = Field()
    prefix = Field()
    suffix = Field()

cust = FixedCustomer()
print(f'Before: {cust.first_name!r} {cust.__dict__}')
cust.first_name = 'Mersenne'
print(f'After:  {cust.first_name!r} {cust.__dict__}')
```

```
>>>
Before: '' {}
After:  'Mersenne' {'_first_name': 'Mersenne'}
```

Things to Remember

+ Metaclasses enable you to modify a class's attributes before the class is fully defined.

+ Descriptors and metaclasses make a powerful combination for declarative behavior and runtime introspection.

+ Define __set_name__ on your descriptor classes to allow them to take into account their surrounding class and its property names.

+ Avoid memory leaks and the weakref built-in module by having descriptors store data they manipulate directly within a class's instance dictionary.

Item 51: Prefer Class Decorators Over Metaclasses for Composable Class Extensions

Although metaclasses allow you to customize class creation in multiple ways (see Item 48: "Validate Subclasses with __init_subclass__" and Item 49: "Register Class Existence with __init_subclass__"), they still fall short of handling every situation that may arise.

For example, say that I want to decorate all of the methods of a class with a helper that prints arguments, return values, and exceptions raised. Here, I define the debugging decorator (see Item 26: "Define Function Decorators with functools.wraps" for background):

```
from functools import wraps

def trace_func(func):
    if hasattr(func, 'tracing'):  # Only decorate once
        return func

    @wraps(func)
    def wrapper(*args, **kwargs):
        result = None
        try:
            result = func(*args, **kwargs)
            return result
        except Exception as e:
            result = e
            raise
```

```
        finally:
            print(f'{func.__name__}({args!r}, {kwargs!r}) -> '
                  f'{result!r}')

    wrapper.tracing = True
    return wrapper
```

I can apply this decorator to various special methods in my new dict
subclass (see Item 43: "Inherit from collections.abc for Custom Container Types" for background):

```
class TraceDict(dict):
    @trace_func
    def __init__(self, *args, **kwargs):
        super().__init__(*args, **kwargs)

    @trace_func
    def __setitem__(self, *args, **kwargs):
        return super().__setitem__(*args, **kwargs)

    @trace_func
    def __getitem__(self, *args, **kwargs):
        return super().__getitem__(*args, **kwargs)

    ...
```

And I can verify that these methods are decorated by interacting with
an instance of the class:

```
trace_dict = TraceDict([('hi', 1)])
trace_dict['there'] = 2
trace_dict['hi']
try:
    trace_dict['does not exist']
except KeyError:
    pass  # Expected
```

```
>>>
__init__(({'hi': 1}, [('hi', 1)]), {}) -> None
__setitem__(({'hi': 1, 'there': 2}, 'there', 2), {}) -> None
__getitem__(({'hi': 1, 'there': 2}, 'hi'), {}) -> 1
__getitem__(({'hi': 1, 'there': 2}, 'does not exist'),
➡{}) -> KeyError('does not exist')
```

The problem with this code is that I had to redefine all of the methods
that I wanted to decorate with @trace_func. This is redundant boilerplate that's hard to read and error prone. Further, if a new method is

later added to the dict superclass, it won't be decorated unless I also define it in TraceDict.

One way to solve this problem is to use a metaclass to automatically decorate all methods of a class. Here, I implement this behavior by wrapping each function or method in the new type with the trace_func decorator:

```python
import types

trace_types = (
    types.MethodType,
    types.FunctionType,
    types.BuiltinFunctionType,
    types.BuiltinMethodType,
    types.MethodDescriptorType,
    types.ClassMethodDescriptorType)


class TraceMeta(type):
    def __new__(meta, name, bases, class_dict):
        klass = super().__new__(meta, name, bases, class_dict)

        for key in dir(klass):
            value = getattr(klass, key)
            if isinstance(value, trace_types):
                wrapped = trace_func(value)
                setattr(klass, key, wrapped)

        return klass
```

Now, I can declare my dict subclass by using the TraceMeta metaclass and verify that it works as expected:

```python
class TraceDict(dict, metaclass=TraceMeta):
    pass


trace_dict = TraceDict([('hi', 1)])
trace_dict['there'] = 2
trace_dict['hi']
try:
    trace_dict['does not exist']
except KeyError:
    pass  # Expected

>>>
__new__((<class '__main__.TraceDict'>, [('hi', 1)]), {}) -> {}
__getitem__(({'hi': 1, 'there': 2}, 'hi'), {}) -> 1
```

```
__getitem__(({'hi': 1, 'there': 2}, 'does not exist'),
➡{}) -> KeyError('does not exist')
```

This works, and it even prints out a call to __new__ that was missing from my earlier implementation. What happens if I try to use TraceMeta when a superclass already has specified a metaclass?

```
class OtherMeta(type):
    pass


class SimpleDict(dict, metaclass=OtherMeta):
    pass


class TraceDict(SimpleDict, metaclass=TraceMeta):
    pass
```

```
>>>
Traceback ...
TypeError: metaclass conflict: the metaclass of a derived
➡class must be a (non-strict) subclass of the metaclasses
➡of all its bases
```

This fails because TraceMeta does not inherit from OtherMeta. In theory, I can use metaclass inheritance to solve this problem by having OtherMeta inherit from TraceMeta:

```
class TraceMeta(type):
    ...


class OtherMeta(TraceMeta):
    pass


class SimpleDict(dict, metaclass=OtherMeta):
    pass


class TraceDict(SimpleDict, metaclass=TraceMeta):
    pass


trace_dict = TraceDict([('hi', 1)])
trace_dict['there'] = 2
trace_dict['hi']
try:
    trace_dict['does not exist']
except KeyError:
    pass  # Expected
```

```
>>>
__init_subclass__((), {}) -> None
__new__((<class '__main__.TraceDict'>, [('hi', 1)]), {}) -> {}
__getitem__(({'hi': 1, 'there': 2}, 'hi'), {}) -> 1
__getitem__(({'hi': 1, 'there': 2}, 'does not exist'),
➡{}) -> KeyError('does not exist')
```

But this won't work if the metaclass is from a library that I can't modify, or if I want to use multiple utility metaclasses like TraceMeta at the same time. The metaclass approach puts too many constraints on the class that's being modified.

To solve this problem, Python supports *class decorators*. Class decorators work just like function decorators: They're applied with the @ symbol prefixing a function before the class declaration. The function is expected to modify or re-create the class accordingly and then return it:

```
def my_class_decorator(klass):
    klass.extra_param = 'hello'
    return klass

@my_class_decorator
class MyClass:
    pass

print(MyClass)
print(MyClass.extra_param)

>>>
<class '__main__.MyClass'>
hello
```

I can implement a class decorator to apply trace_func to all methods and functions of a class by moving the core of the TraceMeta.__new__ method above into a stand-alone function. This implementation is much shorter than the metaclass version:

```
def trace(klass):
    for key in dir(klass):
        value = getattr(klass, key)
        if isinstance(value, trace_types):
            wrapped = trace_func(value)
            setattr(klass, key, wrapped)
    return klass
```

I can apply this decorator to my dict subclass to get the same behavior as I get by using the metaclass approach above:

```
@trace
class TraceDict(dict):
    pass

trace_dict = TraceDict([('hi', 1)])
trace_dict['there'] = 2
trace_dict['hi']
try:
    trace_dict['does not exist']
except KeyError:
    pass  # Expected

>>>
__new__(((<class '__main__.TraceDict'>, [('hi', 1)]), {}) -> {}
__getitem__(({'hi': 1, 'there': 2}, 'hi'), {}) -> 1
__getitem__(({'hi': 1, 'there': 2}, 'does not exist'),
➥{}) -> KeyError('does not exist')
```

Class decorators also work when the class being decorated already has a metaclass:

```
class OtherMeta(type):
    pass

@trace
class TraceDict(dict, metaclass=OtherMeta):
    pass

trace_dict = TraceDict([('hi', 1)])
trace_dict['there'] = 2
trace_dict['hi']
try:
    trace_dict['does not exist']
except KeyError:
    pass  # Expected

>>>
__new__(((<class '__main__.TraceDict'>, [('hi', 1)]), {}) -> {}
__getitem__(({'hi': 1, 'there': 2}, 'hi'), {}) -> 1
__getitem__(({'hi': 1, 'there': 2}, 'does not exist'),
➥{}) -> KeyError('does not exist')
```

When you're looking for composable ways to extend classes, class decorators are the best tool for the job. (See Item 73: "Know How to Use heapq for Priority Queues" for a useful class decorator called functools.total_ordering.)

Things to Remember

✦ A class decorator is a simple function that receives a class instance as a parameter and returns either a new class or a modified version of the original class.

✦ Class decorators are useful when you want to modify every method or attribute of a class with minimal boilerplate.

✦ Metaclasses can't be composed together easily, while many class decorators can be used to extend the same class without conflicts.

7

Chapter

Concurrency and Parallelism

Concurrency enables a computer to do many different things *seemingly* at the same time. For example, on a computer with one CPU core, the operating system rapidly changes which program is running on the single processor. In doing so, it interleaves execution of the programs, providing the illusion that the programs are running simultaneously.

Parallelism, in contrast, involves *actually* doing many different things at the same time. A computer with multiple CPU cores can execute multiple programs simultaneously. Each CPU core runs the instructions of a separate program, allowing each program to make forward progress during the same instant.

Within a single program, concurrency is a tool that makes it easier for programmers to solve certain types of problems. Concurrent programs enable many distinct paths of execution, including separate streams of I/O, to make forward progress in a way that seems to be both simultaneous and independent.

The key difference between parallelism and concurrency is *speedup*. When two distinct paths of execution in a program make forward progress in parallel, the time it takes to do the total work is cut in half; the speed of execution is faster by a factor of two. In contrast, concurrent programs may run thousands of separate paths of execution seemingly in parallel but provide no speedup for the total work.

Python makes it easy to write concurrent programs in a variety of styles. Threads support a relatively small amount of concurrency, while coroutines enable vast numbers of concurrent functions. Python can also be used to do parallel work through system calls, subprocesses, and C extensions. But it can be very difficult to make concurrent Python code truly run in parallel. It's important to understand how to best utilize Python in these different situations.

Item 52: Use subprocess to Manage Child Processes

Python has battle-hardened libraries for running and managing child processes. This makes it a great language for gluing together other tools, such as command-line utilities. When existing shell scripts get complicated, as they often do over time, graduating them to a rewrite in Python for the sake of readability and maintainability is a natural choice.

Child processes started by Python are able to run in parallel, enabling you to use Python to consume all of the CPU cores of a machine and maximize the throughput of programs. Although Python itself may be CPU bound (see Item 53: "Use Threads for Blocking I/O, Avoid for Parallelism"), it's easy to use Python to drive and coordinate CPU-intensive workloads.

Python has many ways to run subprocesses (e.g., os.popen, os.exec*), but the best choice for managing child processes is to use the subprocess built-in module. Running a child process with subprocess is simple. Here, I use the module's run convenience function to start a process, read its output, and verify that it terminated cleanly:

```
import subprocess

result = subprocess.run(
    ['echo', 'Hello from the child!'],
    capture_output=True,
    encoding='utf-8')

result.check_returncode()   # No exception means clean exit
print(result.stdout)

>>>
Hello from the child!
```

Note

> The examples in this item assume that your system has the echo, sleep, and openssl commands available. On Windows, this may not be the case. Please refer to the full example code for this item to see specific directions on how to run these snippets on Windows.

Child processes run independently from their parent process, the Python interpreter. If I create a subprocess using the Popen class instead of the run function, I can poll child process status periodically while Python does other work:

```
proc = subprocess.Popen(['sleep', '1'])
while proc.poll() is None:
    print('Working...')
```

```
    # Some time-consuming work here
    ...

print('Exit status', proc.poll())

>>>
Working...
Working...
Working...
Working...
Exit status 0
```

Decoupling the child process from the parent frees up the parent process to run many child processes in parallel. Here, I do this by starting all the child processes together with Popen upfront:

```
import time

start = time.time()
sleep_procs = []
for _ in range(10):
    proc = subprocess.Popen(['sleep', '1'])
    sleep_procs.append(proc)
```

Later, I wait for them to finish their I/O and terminate with the communicate method:

```
for proc in sleep_procs:
    proc.communicate()

end = time.time()
delta = end - start
print(f'Finished in {delta:.3} seconds')

>>>
Finished in 1.05 seconds
```

If these processes ran in sequence, the total delay would be 10 seconds or more rather than the ~1 second that I measured.

You can also pipe data from a Python program into a subprocess and retrieve its output. This allows you to utilize many other programs to do work in parallel. For example, say that I want to use the openssl command-line tool to encrypt some data. Starting the child process with command-line arguments and I/O pipes is easy:

```
import os
def run_encrypt(data):
    env = os.environ.copy()
```

```
    env['password'] = 'zf7ShyBhZOraQDdE/FiZpm/m/8f9X+M1'
    proc = subprocess.Popen(
        ['openssl', 'enc', '-des3', '-pass', 'env:password'],
        env=env,
        stdin=subprocess.PIPE,
        stdout=subprocess.PIPE)
    proc.stdin.write(data)
    proc.stdin.flush()  # Ensure that the child gets input
    return proc
```

Here, I pipe random bytes into the encryption function, but in practice this input pipe would be fed data from user input, a file handle, a network socket, and so on:

```
procs = []
for _ in range(3):
    data = os.urandom(10)
    proc = run_encrypt(data)
    procs.append(proc)
```

The child processes run in parallel and consume their input. Here, I wait for them to finish and then retrieve their final output. The output is random encrypted bytes as expected:

```
for proc in procs:
    out, _ = proc.communicate()
    print(out[-10:])

>>>
b'\x8c(\xed\xc7m1\xf0F4\xe6'
b'\x0eD\x97\xe9>\x10h{\xbd\xf0'
b'g\x93)\x14U\xa9\xdc\xdd\x04\xd2'
```

It's also possible to create chains of parallel processes, just like UNIX pipelines, connecting the output of one child process to the input of another, and so on. Here's a function that starts the openssl command-line tool as a subprocess to generate a Whirlpool hash of the input stream:

```
def run_hash(input_stdin):
    return subprocess.Popen(
        ['openssl', 'dgst', '-whirlpool', '-binary'],
        stdin=input_stdin,
        stdout=subprocess.PIPE)
```

Now, I can kick off one set of processes to encrypt some data and another set of processes to subsequently hash their encrypted output. Note that I have to be careful with how the stdout instance of the upstream process is retained by the Python interpreter process that's starting this pipeline of child processes:

```
encrypt_procs = []
hash_procs = []
for _ in range(3):
    data = os.urandom(100)

    encrypt_proc = run_encrypt(data)
    encrypt_procs.append(encrypt_proc)

    hash_proc = run_hash(encrypt_proc.stdout)
    hash_procs.append(hash_proc)

    # Ensure that the child consumes the input stream and
    # the communicate() method doesn't inadvertently steal
    # input from the child. Also lets SIGPIPE propagate to
    # the upstream process if the downstream process dies.
    encrypt_proc.stdout.close()
    encrypt_proc.stdout = None
```

The I/O between the child processes happens automatically once they are started. All I need to do is wait for them to finish and print the final output:

```
for proc in encrypt_procs:
    proc.communicate()
    assert proc.returncode == 0

for proc in hash_procs:
    out, _ = proc.communicate()
    print(out[-10:])
    assert proc.returncode == 0
>>>
b'\xe2j\x98h\xfd\xec\xe7T\xd84'
b'\xf3.i\x01\xd74|\xf2\x94E'
b'5_n\xc3-\xe6j\xeb[i'
```

If I'm worried about the child processes never finishing or somehow blocking on input or output pipes, I can pass the timeout parameter to the communicate method. This causes an exception to be raised if the child process hasn't finished within the time period, giving me a chance to terminate the misbehaving subprocess:

```
proc = subprocess.Popen(['sleep', '10'])
try:
    proc.communicate(timeout=0.1)
```

```
except subprocess.TimeoutExpired:
    proc.terminate()
    proc.wait()

print('Exit status', proc.poll())
>>>
Exit status -15
```

Things to Remember

+ Use the subprocess module to run child processes and manage their input and output streams.

+ Child processes run in parallel with the Python interpreter, enabling you to maximize your usage of CPU cores.

+ Use the run convenience function for simple usage, and the Popen class for advanced usage like UNIX-style pipelines.

+ Use the timeout parameter of the communicate method to avoid deadlocks and hanging child processes.

Item 53: Use Threads for Blocking I/O, Avoid for Parallelism

The standard implementation of Python is called CPython. CPython runs a Python program in two steps. First, it parses and compiles the source text into *bytecode*, which is a low-level representation of the program as 8-bit instructions. (As of Python 3.6, however, it's technically *wordcode* with 16-bit instructions, but the idea is the same.) Then, CPython runs the bytecode using a stack-based interpreter. The bytecode interpreter has state that must be maintained and coherent while the Python program executes. CPython enforces coherence with a mechanism called the *global interpreter lock* (GIL).

Essentially, the GIL is a mutual-exclusion lock (mutex) that prevents CPython from being affected by preemptive multithreading, where one thread takes control of a program by interrupting another thread. Such an interruption could corrupt the interpreter state (e.g., garbage collection reference counts) if it comes at an unexpected time. The GIL prevents these interruptions and ensures that every bytecode instruction works correctly with the CPython implementation and its C-extension modules.

The GIL has an important negative side effect. With programs written in languages like C++ or Java, having multiple threads of execution means that a program could utilize multiple CPU cores at the same time. Although Python supports multiple threads of execution, the GIL causes only one of them to ever make forward progress at a time. This means that when you reach for threads to do parallel computation and speed up your Python programs, you will be sorely disappointed.

For example, say that I want to do something computationally intensive with Python. Here, I use a naive number factorization algorithm as a proxy:

```python
def factorize(number):
    for i in range(1, number + 1):
        if number % i == 0:
            yield i
```

Factoring a set of numbers in serial takes quite a long time:

```python
import time

numbers = [2139079, 1214759, 1516637, 1852285]
start = time.time()

for number in numbers:
    list(factorize(number))

end = time.time()
delta = end - start
print(f'Took {delta:.3f} seconds')

>>>
Took 0.399 seconds
```

Using multiple threads to do this computation would make sense in other languages because I could take advantage of all the CPU cores of my computer. Let me try that in Python. Here, I define a Python thread for doing the same computation as before:

```python
from threading import Thread

class FactorizeThread(Thread):
    def __init__(self, number):
        super().__init__()
        self.number = number

    def run(self):
        self.factors = list(factorize(self.number))
```

Then, I start a thread for each number to factorize in parallel:

```
start = time.time()

threads = []
for number in numbers:
    thread = FactorizeThread(number)
    thread.start()
    threads.append(thread)
```

Finally, I wait for all of the threads to finish:

```
for thread in threads:
    thread.join()

end = time.time()
delta = end - start
print(f'Took {delta:.3f} seconds')
```

```
>>>
Took 0.446 seconds
```

Surprisingly, this takes even longer than running factorize in serial. With one thread per number, you might expect less than a 4x speedup in other languages due to the overhead of creating threads and coordinating with them. You might expect only a 2x speedup on the dual-core machine I used to run this code. But you wouldn't expect the performance of these threads to be worse when there are multiple CPUs to utilize. This demonstrates the effect of the GIL (e.g., lock contention and scheduling overhead) on programs running in the standard CPython interpreter.

There are ways to get CPython to utilize multiple cores, but they don't work with the standard Thread class (see Item 64: "Consider concurrent.futures for True Parallelism"), and they can require substantial effort. Given these limitations, why does Python support threads at all? There are two good reasons.

First, multiple threads make it easy for a program to seem like it's doing multiple things at the same time. Managing the juggling act of simultaneous tasks is difficult to implement yourself (see Item 56: "Know How to Recognize When Concurrency Is Necessary" for an example). With threads, you can leave it to Python to run your functions concurrently. This works because CPython ensures a level of fairness between Python threads of execution, even though only one of them makes forward progress at a time due to the GIL.

The second reason Python supports threads is to deal with blocking I/O, which happens when Python does certain types of system calls.

A Python program uses system calls to ask the computer's operating system to interact with the external environment on its behalf. Blocking I/O includes things like reading and writing files, interacting with networks, communicating with devices like displays, and so on. Threads help handle blocking I/O by insulating a program from the time it takes for the operating system to respond to requests.

For example, say that I want to send a signal to a remote-controlled helicopter through a serial port. I'll use a slow system call (select) as a proxy for this activity. This function asks the operating system to block for 0.1 seconds and then return control to my program, which is similar to what would happen when using a synchronous serial port:

```
import select
import socket

def slow_systemcall():
    select.select([socket.socket()], [], [], 0.1)
```

Running this system call in serial requires a linearly increasing amount of time:

```
start = time.time()

for _ in range(5):
    slow_systemcall()

end = time.time()
delta = end - start
print(f'Took {delta:.3f} seconds')

>>>
Took 0.510 seconds
```

The problem is that while the slow_systemcall function is running, my program can't make any other progress. My program's main thread of execution is blocked on the select system call. This situation is awful in practice. You need to be able to compute your helicopter's next move while you're sending it a signal; otherwise, it'll crash. When you find yourself needing to do blocking I/O and computation simultaneously, it's time to consider moving your system calls to threads.

Here, I run multiple invocations of the slow_systemcall function in separate threads. This would allow me to communicate with multiple serial ports (and helicopters) at the same time while leaving the main thread to do whatever computation is required:

```
start = time.time()
```

```
threads = []
for _ in range(5):
    thread = Thread(target=slow_systemcall)
    thread.start()
    threads.append(thread)
```

With the threads started, here I do some work to calculate the next helicopter move before waiting for the system call threads to finish:

```
def compute_helicopter_location(index):
    ...

for i in range(5):
    compute_helicopter_location(i)

for thread in threads:
    thread.join()

end = time.time()
delta = end - start
print(f'Took {delta:.3f} seconds')
```

```
>>>
Took 0.108 seconds
```

The parallel time is ~5x less than the serial time. This shows that all the system calls will run in parallel from multiple Python threads even though they're limited by the GIL. The GIL prevents my Python code from running in parallel, but it doesn't have an effect on system calls. This works because Python threads release the GIL just before they make system calls, and they reacquire the GIL as soon as the system calls are done.

There are many other ways to deal with blocking I/O besides using threads, such as the asyncio built-in module, and these alternatives have important benefits. But those options might require extra work in refactoring your code to fit a different model of execution (see Item 60: "Achieve Highly Concurrent I/O with Coroutines" and Item 62: "Mix Threads and Coroutines to Ease the Transition to asyncio"). Using threads is the simplest way to do blocking I/O in parallel with minimal changes to your program.

Things to Remember

✦ Python threads can't run in parallel on multiple CPU cores because of the global interpreter lock (GIL).

✦ Python threads are still useful despite the GIL because they provide an easy way to do multiple things seemingly at the same time.

✦ Use Python threads to make multiple system calls in parallel. This allows you to do blocking I/O at the same time as computation.

Item 54: Use Lock to Prevent Data Races in Threads

After learning about the global interpreter lock (GIL) (see Item 53: "Use Threads for Blocking I/O, Avoid for Parallelism"), many new Python programmers assume they can forgo using mutual-exclusion locks (also called *mutexes*) in their code altogether. If the GIL is already preventing Python threads from running on multiple CPU cores in parallel, it must also act as a lock for a program's data structures, right? Some testing on types like lists and dictionaries may even show that this assumption appears to hold.

But beware, this is not truly the case. The GIL will not protect you. Although only one Python thread runs at a time, a thread's operations on data structures can be interrupted between any two byte-code instructions in the Python interpreter. This is dangerous if you access the same objects from multiple threads simultaneously. The invariants of your data structures could be violated at practically any time because of these interruptions, leaving your program in a corrupted state.

For example, say that I want to write a program that counts many things in parallel, like sampling light levels from a whole network of sensors. If I want to determine the total number of light samples over time, I can aggregate them with a new class:

```python
class Counter:
    def __init__(self):
        self.count = 0

    def increment(self, offset):
        self.count += offset
```

Imagine that each sensor has its own worker thread because reading from the sensor requires blocking I/O. After each sensor measurement, the worker thread increments the counter up to a maximum number of desired readings:

```python
def worker(sensor_index, how_many, counter):
    for _ in range(how_many):
        # Read from the sensor
        ...
        counter.increment(1)
```

Here, I run one worker thread for each sensor in parallel and wait for them all to finish their readings:

```
from threading import Thread

how_many = 10**5
counter = Counter()

threads = []
for i in range(5):
    thread = Thread(target=worker,
                    args=(i, how_many, counter))
    threads.append(thread)
    thread.start()

for thread in threads:
    thread.join()

expected = how_many * 5
found = counter.count
print(f'Counter should be {expected}, got {found}')

>>>
Counter should be 500000, got 246760
```

This seemed straightforward, and the outcome should have been obvious, but the result is way off! What happened here? How could something so simple go so wrong, especially since only one Python interpreter thread can run at a time?

The Python interpreter enforces fairness between all of the threads that are executing to ensure they get roughly equal processing time. To do this, Python suspends a thread as it's running and resumes another thread in turn. The problem is that you don't know exactly when Python will suspend your threads. A thread can even be paused seemingly halfway through what looks like an atomic operation. That's what happened in this case.

The body of the Counter object's increment method looks simple, and is equivalent to this statement from the perspective of the worker thread:

```
counter.count += 1
```

But the += operator used on an object attribute actually instructs Python to do three separate operations behind the scenes. The statement above is equivalent to this:

```
value = getattr(counter, 'count')
result = value + 1
setattr(counter, 'count', result)
```

Python threads incrementing the counter can be suspended between any two of these operations. This is problematic if the way the operations interleave causes old versions of value to be assigned to the counter. Here's an example of bad interaction between two threads, A and B:

```
# Running in Thread A
value_a = getattr(counter, 'count')
# Context switch to Thread B
value_b = getattr(counter, 'count')
result_b = value_b + 1
setattr(counter, 'count', result_b)
# Context switch back to Thread A
result_a = value_a + 1
setattr(counter, 'count', result_a)
```

Thread B interrupted thread A before it had completely finished. Thread B ran and finished, but then thread A resumed mid-execution, overwriting all of thread B's progress in incrementing the counter. This is exactly what happened in the light sensor example above.

To prevent data races like these, and other forms of data structure corruption, Python includes a robust set of tools in the threading built-in module. The simplest and most useful of them is the Lock class, a mutual-exclusion lock (mutex).

By using a lock, I can have the Counter class protect its current value against simultaneous accesses from multiple threads. Only one thread will be able to acquire the lock at a time. Here, I use a with statement to acquire and release the lock; this makes it easier to see which code is executing while the lock is held (see Item 66: "Consider contextlib and with Statements for Reusable try/finally Behavior" for background):

```
from threading import Lock

class LockingCounter:
    def __init__(self):
        self.lock = Lock()
        self.count = 0

    def increment(self, offset):
        with self.lock:
            self.count += offset
```

Now, I run the worker threads as before but use a LockingCounter instead:

```
counter = LockingCounter()

for i in range(5):
    thread = Thread(target=worker,
                        args=(i, how_many, counter))
    threads.append(thread)
    thread.start()

for thread in threads:
    thread.join()

expected = how_many * 5
found = counter.count
print(f'Counter should be {expected}, got {found}')
>>>
Counter should be 500000, got 500000
```

The result is exactly what I expect. Lock solved the problem.

Things to Remember

✦ Even though Python has a global interpreter lock, you're still responsible for protecting against data races between the threads in your programs.

✦ Your programs will corrupt their data structures if you allow multiple threads to modify the same objects without mutual-exclusion locks (mutexes).

✦ Use the Lock class from the threading built-in module to enforce your program's invariants between multiple threads.

Item 55: Use Queue to Coordinate Work Between Threads

Python programs that do many things concurrently often need to coordinate their work. One of the most useful arrangements for concurrent work is a pipeline of functions.

A pipeline works like an assembly line used in manufacturing. Pipelines have many phases in serial, with a specific function for each phase. New pieces of work are constantly being added to the beginning of the pipeline. The functions can operate concurrently, each working on the piece of work in its phase. The work moves forward as each function completes until there are no phases remaining. This approach is especially good for work that includes blocking I/O or subprocesses—activities that can easily be parallelized using Python (see Item 53: "Use Threads for Blocking I/O, Avoid for Parallelism").

For example, say I want to build a system that will take a constant stream of images from my digital camera, resize them, and then add them to a photo gallery online. Such a program could be split into three phases of a pipeline. New images are retrieved in the first phase. The downloaded images are passed through the resize function in the second phase. The resized images are consumed by the upload function in the final phase.

Imagine that I've already written Python functions that execute the phases: download, resize, upload. How do I assemble a pipeline to do the work concurrently?

```python
def download(item):
    ...

def resize(item):
    ...

def upload(item):
    ...
```

The first thing I need is a way to hand off work between the pipeline phases. This can be modeled as a thread-safe producer–consumer queue (see Item 54: "Use Lock to Prevent Data Races in Threads" to understand the importance of thread safety in Python; see Item 71: "Prefer deque for Producer–Consumer Queues" to understand queue performance):

```python
from collections import deque
from threading import Lock

class MyQueue:
    def __init__(self):
        self.items = deque()
        self.lock = Lock()
```

The producer, my digital camera, adds new images to the end of the deque of pending items:

```python
    def put(self, item):
        with self.lock:
            self.items.append(item)
```

The consumer, the first phase of the processing pipeline, removes images from the front of the deque of pending items:

```
def get(self):
    with self.lock:
        return self.items.popleft()
```

Here, I represent each phase of the pipeline as a Python thread that takes work from one queue like this, runs a function on it, and puts the result on another queue. I also track how many times the worker has checked for new input and how much work it's completed:

```
from threading import Thread
import time

class Worker(Thread):
    def __init__(self, func, in_queue, out_queue):
        super().__init__()
        self.func = func
        self.in_queue = in_queue
        self.out_queue = out_queue
        self.polled_count = 0
        self.work_done = 0
```

The trickiest part is that the worker thread must properly handle the case where the input queue is empty because the previous phase hasn't completed its work yet. This happens where I catch the IndexError exception below. You can think of this as a holdup in the assembly line:

```
def run(self):
    while True:
        self.polled_count += 1
        try:
            item = self.in_queue.get()
        except IndexError:
            time.sleep(0.01)  # No work to do
        else:
            result = self.func(item)
            self.out_queue.put(result)
            self.work_done += 1
```

Now, I can connect the three phases together by creating the queues for their coordination points and the corresponding worker threads:

```
download_queue = MyQueue()
resize_queue = MyQueue()
upload_queue = MyQueue()
```

```
done_queue = MyQueue()
threads = [
    Worker(download, download_queue, resize_queue),
    Worker(resize, resize_queue, upload_queue),
    Worker(upload, upload_queue, done_queue),
]
```

I can start the threads and then inject a bunch of work into the first phase of the pipeline. Here, I use a plain object instance as a proxy for the real data required by the download function:

```
for thread in threads:
    thread.start()

for _ in range(1000):
    download_queue.put(object())
```

Now, I wait for all of the items to be processed by the pipeline and end up in the done_queue:

```
while len(done_queue.items) < 1000:
    # Do something useful while waiting
    ...
```

This runs properly, but there's an interesting side effect caused by the threads polling their input queues for new work. The tricky part, where I catch IndexError exceptions in the run method, executes a large number of times:

```
processed = len(done_queue.items)
polled = sum(t.polled_count for t in threads)
print(f'Processed {processed} items after '
      f'polling {polled} times')
```

```
>>>
Processed 1000 items after polling 3035 times
```

When the worker functions vary in their respective speeds, an earlier phase can prevent progress in later phases, backing up the pipeline. This causes later phases to starve and constantly check their input queues for new work in a tight loop. The outcome is that worker threads waste CPU time doing nothing useful; they're constantly raising and catching IndexError exceptions.

But that's just the beginning of what's wrong with this implementation. There are three more problems that you should also avoid. First, determining that all of the input work is complete requires yet another busy wait on the done_queue. Second, in Worker, the run method will execute forever in its busy loop. There's no obvious way to signal to a worker thread that it's time to exit.

Third, and worst of all, a backup in the pipeline can cause the program to crash arbitrarily. If the first phase makes rapid progress but the second phase makes slow progress, then the queue connecting the first phase to the second phase will constantly increase in size. The second phase won't be able to keep up. Given enough time and input data, the program will eventually run out of memory and die.

The lesson here isn't that pipelines are bad; it's that it's hard to build a good producer–consumer queue yourself. So why even try?

Queue to the Rescue

The Queue class from the queue built-in module provides all of the functionality you need to solve the problems outlined above.

Queue eliminates the busy waiting in the worker by making the get method block until new data is available. For example, here I start a thread that waits for some input data on a queue:

```
from queue import Queue

my_queue = Queue()

def consumer():
    print('Consumer waiting')
    my_queue.get()                   # Runs after put() below
    print('Consumer done')

thread = Thread(target=consumer)
thread.start()
```

Even though the thread is running first, it won't finish until an item is put on the Queue instance and the get method has something to return:

```
print('Producer putting')
my_queue.put(object())               # Runs before get() above
print('Producer done')
thread.join()

>>>
Consumer waiting
Producer putting
Producer done
Consumer done
```

To solve the pipeline backup issue, the Queue class lets you specify the maximum amount of pending work to allow between two phases.

This buffer size causes calls to put to block when the queue is already full. For example, here I define a thread that waits for a while before consuming a queue:

```
my_queue = Queue(1)                  # Buffer size of 1

def consumer():
    time.sleep(0.1)                  # Wait
    my_queue.get()                   # Runs second
    print('Consumer got 1')
    my_queue.get()                   # Runs fourth
    print('Consumer got 2')
    print('Consumer done')

thread = Thread(target=consumer)
thread.start()
```

The wait should allow the producer thread to put both objects on the queue before the consumer thread ever calls get. But the Queue size is one. This means the producer adding items to the queue will have to wait for the consumer thread to call get at least once before the second call to put will stop blocking and add the second item to the queue:

```
my_queue.put(object())               # Runs first
print('Producer put 1')
my_queue.put(object())               # Runs third
print('Producer put 2')
print('Producer done')
thread.join()
```

```
>>>
Producer put 1
Consumer got 1
Producer put 2
Producer done
Consumer got 2
Consumer done
```

The Queue class can also track the progress of work using the task_done method. This lets you wait for a phase's input queue to drain and eliminates the need to poll the last phase of a pipeline (as with the done_queue above). For example, here I define a consumer thread that calls task_done when it finishes working on an item:

```
in_queue = Queue()
```

```
def consumer():
    print('Consumer waiting')
    work = in_queue.get()         # Runs second
    print('Consumer working')
    # Doing work
    ...
    print('Consumer done')
    in_queue.task_done()          # Runs third

thread = Thread(target=consumer)
thread.start()
```

Now, the producer code doesn't have to join the consumer thread or poll. The producer can just wait for the in_queue to finish by calling join on the Queue instance. Even once it's empty, the in_queue won't be joinable until after task_done is called for every item that was ever enqueued:

```
print('Producer putting')
in_queue.put(object())            # Runs first
print('Producer waiting')
in_queue.join()                   # Runs fourth
print('Producer done')
thread.join()

>>>
Consumer waiting
Producer putting
Producer waiting
Consumer working
Consumer done
Producer done
```

I can put all these behaviors together into a Queue subclass that also tells the worker thread when it should stop processing. Here, I define a close method that adds a special *sentinel* item to the queue that indicates there will be no more input items after it:

```
class ClosableQueue(Queue):
    SENTINEL = object()

    def close(self):
        self.put(self.SENTINEL)
```

Then, I define an iterator for the queue that looks for this special object and stops iteration when it's found. This __iter__ method also calls task_done at appropriate times, letting me track the progress of work on the queue (see Item 31: "Be Defensive When Iterating Over Arguments" for details about __iter__):

```
def __iter__(self):
    while True:
        item = self.get()
        try:
            if item is self.SENTINEL:
                return  # Cause the thread to exit
            yield item
        finally:
            self.task_done()
```

Now, I can redefine my worker thread to rely on the behavior of the ClosableQueue class. The thread will exit when the for loop is exhausted:

```
class StoppableWorker(Thread):
    def __init__(self, func, in_queue, out_queue):
        super().__init__()
        self.func = func
        self.in_queue = in_queue
        self.out_queue = out_queue

    def run(self):
        for item in self.in_queue:
            result = self.func(item)
            self.out_queue.put(result)
```

I re-create the set of worker threads using the new worker class:

```
download_queue = ClosableQueue()
resize_queue = ClosableQueue()
upload_queue = ClosableQueue()
done_queue = ClosableQueue()
threads = [
    StoppableWorker(download, download_queue, resize_queue),
    StoppableWorker(resize, resize_queue, upload_queue),
    StoppableWorker(upload, upload_queue, done_queue),
]
```

After running the worker threads as before, I also send the stop signal after all the input work has been injected by closing the input queue of the first phase:

```
for thread in threads:
    thread.start()
```

```
for _ in range(1000):
    download_queue.put(object())

download_queue.close()
```

Finally, I wait for the work to finish by joining the queues that connect the phases. Each time one phase is done, I signal the next phase to stop by closing its input queue. At the end, the done_queue contains all of the output objects, as expected:

```
download_queue.join()
resize_queue.close()
resize_queue.join()
upload_queue.close()
upload_queue.join()
print(done_queue.qsize(), 'items finished')

for thread in threads:
    thread.join()

>>>
1000 items finished
```

This approach can be extended to use multiple worker threads per phase, which can increase I/O parallelism and speed up this type of program significantly. To do this, first I define some helper functions that start and stop multiple threads. The way stop_threads works is by calling close on each input queue once per consuming thread, which ensures that all of the workers exit cleanly:

```
def start_threads(count, *args):
    threads = [StoppableWorker(*args) for _ in range(count)]
    for thread in threads:
        thread.start()
    return threads

def stop_threads(closable_queue, threads):
    for _ in threads:
        closable_queue.close()

    closable_queue.join()

    for thread in threads:
        thread.join()
```

Then, I connect the pieces together as before, putting objects to process into the top of the pipeline, joining queues and threads along the way, and finally consuming the results:

```
download_queue = ClosableQueue()
resize_queue = ClosableQueue()
upload_queue = ClosableQueue()
done_queue = ClosableQueue()

download_threads = start_threads(
    3, download, download_queue, resize_queue)
resize_threads = start_threads(
    4, resize, resize_queue, upload_queue)
upload_threads = start_threads(
    5, upload, upload_queue, done_queue)

for _ in range(1000):
    download_queue.put(object())

stop_threads(download_queue, download_threads)
stop_threads(resize_queue, resize_threads)
stop_threads(upload_queue, upload_threads)

print(done_queue.qsize(), 'items finished')

>>>
1000 items finished
```

Although Queue works well in this case of a linear pipeline, there are many other situations for which there are better tools that you should consider (see Item 60: "Achieve Highly Concurrent I/O with Coroutines").

Things to Remember

✦ Pipelines are a great way to organize sequences of work—especially I/O-bound programs—that run concurrently using multiple Python threads.

✦ Be aware of the many problems in building concurrent pipelines: busy waiting, how to tell workers to stop, and potential memory explosion.

✦ The Queue class has all the facilities you need to build robust pipelines: blocking operations, buffer sizes, and joining.

Item 56: Know How to Recognize When Concurrency Is Necessary

Inevitably, as the scope of a program grows, it also becomes more complicated. Dealing with expanding requirements in a way that maintains clarity, testability, and efficiency is one of the most difficult parts of programming. Perhaps the hardest type of change to handle is moving from a single-threaded program to one that needs multiple concurrent lines of execution.

Let me demonstrate how you might encounter this problem with an example. Say that I want to implement Conway's Game of Life, a classic illustration of finite state automata. The rules of the game are simple: You have a two-dimensional grid of an arbitrary size. Each cell in the grid can either be alive or empty:

```
ALIVE = '*'
EMPTY = '-'
```

The game progresses one tick of the clock at a time. Every tick, each cell counts how many of its neighboring eight cells are still alive. Based on its neighbor count, a cell decides if it will keep living, die, or regenerate. (I'll explain the specific rules further below.) Here's an example of a 5 × 5 Game of Life grid after four generations with time going to the right:

```
  0   |   1   |   2   |   3   |   4
----- | ----- | ----- | ----- | -----
-*--- | --*-- | --**- | --*-- | -----
--**- | --**- | -*--- | -*--- | -**--
---*- | --**- | --**- | --*-- | -----
----- | ----- | ----- | ----- | -----
```

I can represent the state of each cell with a simple container class. The class must have methods that allow me to get and set the value of any coordinate. Coordinates that are out of bounds should wrap around, making the grid act like an infinite looping space:

```
class Grid:
    def __init__(self, height, width):
        self.height = height
        self.width = width
        self.rows = []
        for _ in range(self.height):
            self.rows.append([EMPTY] * self.width)
```

```
def get(self, y, x):
    return self.rows[y % self.height][x % self.width]

def set(self, y, x, state):
    self.rows[y % self.height][x % self.width] = state

def __str__(self):
    ...
```

To see this class in action, I can create a Grid instance and set its initial state to a classic shape called a glider:

```
grid = Grid(5, 9)
grid.set(0, 3, ALIVE)
grid.set(1, 4, ALIVE)
grid.set(2, 2, ALIVE)
grid.set(2, 3, ALIVE)
grid.set(2, 4, ALIVE)
print(grid)

>>>
---*-----
----*----
--***----
---------
---------
```

Now, I need a way to retrieve the status of neighboring cells. I can do this with a helper function that queries the grid and returns the count of living neighbors. I use a simple function for the get parameter instead of passing in a whole Grid instance in order to reduce coupling (see Item 38: "Accept Functions Instead of Classes for Simple Interfaces" for more about this approach):

```
def count_neighbors(y, x, get):
    n_ = get(y - 1, x + 0)  # North
    ne = get(y - 1, x + 1)  # Northeast
    e_ = get(y + 0, x + 1)  # East
    se = get(y + 1, x + 1)  # Southeast
    s_ = get(y + 1, x + 0)  # South
    sw = get(y + 1, x - 1)  # Southwest
    w_ = get(y + 0, x - 1)  # West
    nw = get(y - 1, x - 1)  # Northwest
    neighbor_states = [n_, ne, e_, se, s_, sw, w_, nw]
    count = 0
```

```
    for state in neighbor_states:
        if state == ALIVE:
            count += 1
    return count
```

Now, I define the simple logic for Conway's Game of Life, based on the game's three rules: Die if a cell has fewer than two neighbors, die if a cell has more than three neighbors, or become alive if an empty cell has exactly three neighbors:

```
def game_logic(state, neighbors):
    if state == ALIVE:
        if neighbors < 2:
            return EMPTY      # Die: Too few
        elif neighbors > 3:
            return EMPTY      # Die: Too many
    else:
        if neighbors == 3:
            return ALIVE      # Regenerate
    return state
```

I can connect count_neighbors and game_logic together in another function that transitions the state of a cell. This function will be called each generation to figure out a cell's current state, inspect the neighboring cells around it, determine what its next state should be, and update the resulting grid accordingly. Again, I use a function interface for set instead of passing in the Grid instance to make this code more decoupled:

```
def step_cell(y, x, get, set):
    state = get(y, x)
    neighbors = count_neighbors(y, x, get)
    next_state = game_logic(state, neighbors)
    set(y, x, next_state)
```

Finally, I can define a function that progresses the whole grid of cells forward by a single step and then returns a new grid containing the state for the next generation. The important detail here is that I need all dependent functions to call the get method on the previous generation's Grid instance, and to call the set method on the next generation's Grid instance. This is how I ensure that all of the cells move in lockstep, which is an essential part of how the game works. This is easy to achieve because I used function interfaces for get and set instead of passing Grid instances:

```
def simulate(grid):
    next_grid = Grid(grid.height, grid.width)
```

```
    for y in range(grid.height):
        for x in range(grid.width):
            step_cell(y, x, grid.get, next_grid.set)
    return next_grid
```

Now, I can progress the grid forward one generation at a time. You can see how the glider moves down and to the right on the grid based on the simple rules from the game_logic function:

```
class ColumnPrinter:
    ...


columns = ColumnPrinter()
for i in range(5):
    columns.append(str(grid))
    grid = simulate(grid)


print(columns)
```

```
>>>
    0     |     1     |     2     |     3     |     4
---*----- | --------- | --------- | --------- | ---------
----*---- | --*-*---- | ----*---- | ---*----- | ----*----
--***---- | ---**---- | --*-*---- | ----**--- | -----*---
--------- | ---*----- | ---**---- | ---**---- | ---***---
--------- | --------- | --------- | --------- | ---------
```

This works great for a program that can run in one thread on a single machine. But imagine that the program's requirements have changed—as I alluded to above—and now I need to do some I/O (e.g., with a socket) from within the game_logic function. For example, this might be required if I'm trying to build a massively multiplayer online game where the state transitions are determined by a combination of the grid state and communication with other players over the Internet.

How can I extend this implementation to support such functionality? The simplest thing to do is to add blocking I/O directly into the game_logic function:

```
def game_logic(state, neighbors):
    ...
    # Do some blocking input/output in here:
    data = my_socket.recv(100)
    ...
```

The problem with this approach is that it's going to slow down the whole program. If the latency of the I/O required is 100 milliseconds (i.e., a reasonably good cross-country, round-trip latency on the

Internet), and there are 45 cells in the grid, then each generation will take a minimum of 4.5 seconds to evaluate because each cell is processed serially in the simulate function. That's far too slow and will make the game unplayable. It also scales poorly: If I later wanted to expand the grid to 10,000 cells, I would need over 15 minutes to evaluate each generation.

The solution is to do the I/O in parallel so each generation takes roughly 100 milliseconds, regardless of how big the grid is. The process of spawning a concurrent line of execution for each unit of work—a cell in this case—is called *fan-out*. Waiting for all of those concurrent units of work to finish before moving on to the next phase in a coordinated process—a generation in this case—is called *fan-in*.

Python provides many built-in tools for achieving fan-out and fan-in with various trade-offs. You should understand the pros and cons of each approach and choose the best tool for the job, depending on the situation. See the items that follow for details based on this Game of Life example program (Item 57: "Avoid Creating New Thread Instances for On-demand Fan-out," Item 58: "Understand How Using Queue for Concurrency Requires Refactoring," Item 59: "Consider ThreadPoolExecutor When Threads Are Necessary for Concurrency," and Item 60: "Achieve Highly Concurrent I/O with Coroutines").

Things to Remember

✦ A program often grows to require multiple concurrent lines of execution as its scope and complexity increases.

✦ The most common types of concurrency coordination are fan-out (generating new units of concurrency) and fan-in (waiting for existing units of concurrency to complete).

✦ Python has many different ways of achieving fan-out and fan-in.

Item 57: Avoid Creating New Thread Instances for On-demand Fan-out

Threads are the natural first tool to reach for in order to do parallel I/O in Python (see Item 53: "Use Threads for Blocking I/O, Avoid for Parallelism"). However, they have significant downsides when you try to use them for fanning out to many concurrent lines of execution.

To demonstrate this, I'll continue with the Game of Life example from before (see Item 56: "Know How to Recognize When Concurrency Is Necessary" for background and the implementations of various functions and classes below). I'll use threads to solve the latency problem caused by doing I/O in the game_logic function. To begin, threads require coordination using locks to ensure that assumptions within data structures are maintained properly. I can create a subclass of the Grid class that adds locking behavior so an instance can be used by multiple threads simultaneously:

```
from threading import Lock

ALIVE = '*'
EMPTY = '-'

class Grid:
    ...

class LockingGrid(Grid):
    def __init__(self, height, width):
        super().__init__(height, width)
        self.lock = Lock()

    def __str__(self):
        with self.lock:
            return super().__str__()

    def get(self, y, x):
        with self.lock:
            return super().get(y, x)

    def set(self, y, x, state):
        with self.lock:
            return super().set(y, x, state)
```

Then, I can reimplement the simulate function to *fan out* by creating a thread for each call to step_cell. The threads will run in parallel and won't have to wait on each other's I/O. I can then *fan in* by waiting for all of the threads to complete before moving on to the next generation:

```
from threading import Thread

def count_neighbors(y, x, get):
    ...

def game_logic(state, neighbors):
    ...
    # Do some blocking input/output in here:
    data = my_socket.recv(100)
    ...
```

```
def step_cell(y, x, get, set):
    state = get(y, x)
    neighbors = count_neighbors(y, x, get)
    next_state = game_logic(state, neighbors)
    set(y, x, next_state)

def simulate_threaded(grid):
    next_grid = LockingGrid(grid.height, grid.width)

    threads = []
    for y in range(grid.height):
        for x in range(grid.width):
            args = (y, x, grid.get, next_grid.set)
            thread = Thread(target=step_cell, args=args)
            thread.start()  # Fan out
            threads.append(thread)

    for thread in threads:
        thread.join()        # Fan in

    return next_grid
```

I can run this code using the same implementation of step_cell and the same driving code as before with only two lines changed to use the LockingGrid and simulate_threaded implementations:

```
class ColumnPrinter:
    ...

grid = LockingGrid(5, 9)              # Changed
grid.set(0, 3, ALIVE)
grid.set(1, 4, ALIVE)
grid.set(2, 2, ALIVE)
grid.set(2, 3, ALIVE)
grid.set(2, 4, ALIVE)

columns = ColumnPrinter()
for i in range(5):
    columns.append(str(grid))
    grid = simulate_threaded(grid)  # Changed

print(columns)
```

```
>>>
    0     |     1     |     2     |     3     |     4
---*----- | --------- | --------- | --------- | ---------
----*---- | --*-*---- | ----*---- | ---*----- | ----*----
--***---- | ---**---- | --*-*---- | ----**--- | -----*---
--------- | ---*----- | ---**---- | ---**---- | ---***---
--------- | --------- | --------- | --------- | ---------
```

This works as expected, and the I/O is now parallelized between the threads. However, this code has three big problems:

- The Thread instances require special tools to coordinate with each other safely (see Item 54: "Use Lock to Prevent Data Races in Threads"). This makes the code that uses threads harder to reason about than the procedural, single-threaded code from before. This complexity makes threaded code more difficult to extend and maintain over time.

- Threads require a lot of memory—about 8 MB per executing thread. On many computers, that amount of memory doesn't matter for the 45 threads I'd need in this example. But if the game grid had to grow to 10,000 cells, I would need to create that many threads, which couldn't even fit in the memory of my machine. Running a thread per concurrent activity just won't work.

- Starting a thread is costly, and threads have a negative performance impact when they run due to context switching between them. In this case, all of the threads are started and stopped each generation of the game, which has high overhead and will increase latency beyond the expected I/O time of 100 milliseconds.

This code would also be very difficult to debug if something went wrong. For example, imagine that the game_logic function raises an exception, which is highly likely due to the generally flaky nature of I/O:

```python
def game_logic(state, neighbors):
    ...
    raise OSError('Problem with I/O')
    ...
```

I can test what this would do by running a Thread instance pointed at this function and redirecting the sys.stderr output from the program to an in-memory StringIO buffer:

```python
import contextlib
import io
```

```
fake_stderr = io.StringIO()
with contextlib.redirect_stderr(fake_stderr):
    thread = Thread(target=game_logic, args=(ALIVE, 3))
    thread.start()
    thread.join()

print(fake_stderr.getvalue())

>>>
Exception in thread Thread-226:
Traceback (most recent call last):
  File "threading.py", line 917, in _bootstrap_inner
    self.run()
  File "threading.py", line 865, in run
    self._target(*self._args, **self._kwargs)
  File "example.py", line 193, in game_logic
    raise OSError('Problem with I/O')
OSError: Problem with I/O
```

An OSError exception is raised as expected, but somehow the code that created the Thread and called join on it is unaffected. How can this be? The reason is that the Thread class will independently catch any exceptions that are raised by the target function and then write their traceback to sys.stderr. Such exceptions are never re-raised to the caller that started the thread in the first place.

Given all of these issues, it's clear that threads are not the solution if you need to constantly create and finish new concurrent functions. Python provides other solutions that are a better fit (see Item 58: "Understand How Using Queue for Concurrency Requires Refactoring," Item 59: "Consider ThreadPoolExecutor When Threads Are Necessary for Concurrency", and Item 60: "Achieve Highly Concurrent I/O with Coroutines").

Things to Remember

✦ Threads have many downsides: They're costly to start and run if you need a lot of them, they each require a significant amount of memory, and they require special tools like Lock instances for coordination.

✦ Threads do not provide a built-in way to raise exceptions back in the code that started a thread or that is waiting for one to finish, which makes them difficult to debug.

Item 58: Understand How Using Queue for Concurrency Requires Refactoring

In the previous item (see Item 57: "Avoid Creating New Thread Instances for On-demand Fan-out") I covered the downsides of using Thread to solve the parallel I/O problem in the Game of Life example from earlier (see Item 56: "Know How to Recognize When Concurrency Is Necessary" for background and the implementations of various functions and classes below).

The next approach to try is to implement a threaded pipeline using the Queue class from the queue built-in module (see Item 55: "Use Queue to Coordinate Work Between Threads" for background; I rely on the implementations of ClosableQueue and StoppableWorker from that item in the example code below).

Here's the general approach: Instead of creating one thread per cell per generation of the Game of Life, I can create a fixed number of worker threads upfront and have them do parallelized I/O as needed. This will keep my resource usage under control and eliminate the overhead of frequently starting new threads.

To do this, I need two ClosableQueue instances to use for communicating to and from the worker threads that execute the game_logic function:

```
from queue import Queue

class ClosableQueue(Queue):
    ...

in_queue = ClosableQueue()
out_queue = ClosableQueue()
```

I can start multiple threads that will consume items from the in_queue, process them by calling game_logic, and put the results on out_queue. These threads will run concurrently, allowing for parallel I/O and reduced latency for each generation:

```
from threading import Thread

class StoppableWorker(Thread):
    ...

def game_logic(state, neighbors):
    ...
```

```
    # Do some blocking input/output in here:
    data = my_socket.recv(100)
    ...

def game_logic_thread(item):
    y, x, state, neighbors = item
    try:
        next_state = game_logic(state, neighbors)
    except Exception as e:
        next_state = e
    return (y, x, next_state)

# Start the threads upfront
threads = []
for _ in range(5):
    thread = StoppableWorker(
        game_logic_thread, in_queue, out_queue)
    thread.start()
    threads.append(thread)
```

Now, I can redefine the simulate function to interact with these queues to request state transition decisions and receive corresponding responses. Adding items to in_queue causes *fan-out*, and consuming items from out_queue until it's empty causes *fan-in*:

```
ALIVE = '*'
EMPTY = '-'

class SimulationError(Exception):
    pass

class Grid:
    ...

def count_neighbors(y, x, get):
    ...

def simulate_pipeline(grid, in_queue, out_queue):
    for y in range(grid.height):
        for x in range(grid.width):
            state = grid.get(y, x)
            neighbors = count_neighbors(y, x, grid.get)
            in_queue.put((y, x, state, neighbors))  # Fan out

    in_queue.join()
    out_queue.close()
```

```
    next_grid = Grid(grid.height, grid.width)
    for item in out_queue:                          # Fan in
        y, x, next_state = item
        if isinstance(next_state, Exception):
            raise SimulationError(y, x) from next_state
        next_grid.set(y, x, next_state)

    return next_grid
```

The calls to Grid.get and Grid.set both happen within this new
simulate_pipeline function, which means I can use the single-threaded
implementation of Grid instead of the implementation that requires
Lock instances for synchronization.

This code is also easier to debug than the Thread approach used
in the previous item. If an exception occurs while doing I/O in the
game_logic function, it will be caught, propagated to the out_queue,
and then re-raised in the main thread:

```
def game_logic(state, neighbors):
    ...
    raise OSError('Problem with I/O in game_logic')
    ...

simulate_pipeline(Grid(1, 1), in_queue, out_queue)
>>>
Traceback ...
OSError: Problem with I/O in game_logic
```

The above exception was the direct cause of the following
➡exception:

```
Traceback ...
SimulationError: (0, 0)
```

I can drive this multithreaded pipeline for repeated generations by
calling simulate_pipeline in a loop:

```
class ColumnPrinter:
    ...

grid = Grid(5, 9)
grid.set(0, 3, ALIVE)
grid.set(1, 4, ALIVE)
grid.set(2, 2, ALIVE)
grid.set(2, 3, ALIVE)
```

```
grid.set(2, 4, ALIVE)

columns = ColumnPrinter()
for i in range(5):
    columns.append(str(grid))
    grid = simulate_pipeline(grid, in_queue, out_queue)

print(columns)

for thread in threads:
    in_queue.close()
for thread in threads:
    thread.join()
```

```
>>>
    0        |     1       |     2       |     3       |     4
---*_____   | ---------   | ---------   | ---------   | ---------
----*____   | ---------   | --*_*____   | ---------   | ----*____
--***____   | ---------   | ---**____   | ---------   | --*_*____
---------   | ---------   | ---*_____   | ---------   | ---**____
---------   | ---------   | ---------   | ---------   | ---------
```

The results are the same as before. Although I've addressed the memory explosion problem, startup costs, and debugging issues of using threads on their own, many issues remain:

- The `simulate_pipeline` function is even harder to follow than the `simulate_threaded` approach from the previous item.

- Extra support classes were required for `ClosableQueue` and `StoppableWorker` in order to make the code easier to read, at the expense of increased complexity.

- I have to specify the amount of potential parallelism—the number of threads running `game_logic_thread`—upfront based on my expectations of the workload instead of having the system automatically scale up parallelism as needed.

- In order to enable debugging, I have to manually catch exceptions in worker threads, propagate them on a `Queue`, and then re-raise them in the main thread.

However, the biggest problem with this code is apparent if the requirements change again. Imagine that later I needed to do I/O within the `count_neighbors` function in addition to the I/O that was needed within `game_logic`:

```
def count_neighbors(y, x, get):
    ...
```

```
    # Do some blocking input/output in here:
    data = my_socket.recv(100)
    ...
```

In order to make this parallelizable, I need to add another stage to the pipeline that runs count_neighbors in a thread. I need to make sure that exceptions propagate correctly between the worker threads and the main thread. And I need to use a Lock for the Grid class in order to ensure safe synchronization between the worker threads (see Item 54: "Use Lock to Prevent Data Races in Threads" for background and Item 57: "Avoid Creating New Thread Instances for On-demand Fan-out" for the implementation of LockingGrid):

```
def count_neighbors_thread(item):
    y, x, state, get = item
    try:
        neighbors = count_neighbors(y, x, get)
    except Exception as e:
        neighbors = e
    return (y, x, state, neighbors)

def game_logic_thread(item):
    y, x, state, neighbors = item
    if isinstance(neighbors, Exception):
        next_state = neighbors
    else:
        try:
            next_state = game_logic(state, neighbors)
        except Exception as e:
            next_state = e
    return (y, x, next_state)

class LockingGrid(Grid):
    ...
```

I have to create another set of Queue instances for the count_neighbors_thread workers and the corresponding Thread instances:

```
in_queue = ClosableQueue()
logic_queue = ClosableQueue()
out_queue = ClosableQueue()

threads = []
```

```
for _ in range(5):
    thread = StoppableWorker(
        count_neighbors_thread, in_queue, logic_queue)
    thread.start()
    threads.append(thread)

for _ in range(5):
    thread = StoppableWorker(
        game_logic_thread, logic_queue, out_queue)
    thread.start()
    threads.append(thread)
```

Finally, I need to update `simulate_pipeline` to coordinate the multiple phases in the pipeline and ensure that work fans out and back in correctly:

```
def simulate_phased_pipeline(
        grid, in_queue, logic_queue, out_queue):
    for y in range(grid.height):
        for x in range(grid.width):
            state = grid.get(y, x)
            item = (y, x, state, grid.get)
            in_queue.put(item)          # Fan out

    in_queue.join()
    logic_queue.join()                  # Pipeline sequencing
    out_queue.close()

    next_grid = LockingGrid(grid.height, grid.width)
    for item in out_queue:              # Fan in
        y, x, next_state = item
        if isinstance(next_state, Exception):
            raise SimulationError(y, x) from next_state
        next_grid.set(y, x, next_state)

    return next_grid
```

With these updated implementations, now I can run the multiphase pipeline end-to-end:

```
grid = LockingGrid(5, 9)
grid.set(0, 3, ALIVE)
grid.set(1, 4, ALIVE)
grid.set(2, 2, ALIVE)
grid.set(2, 3, ALIVE)
grid.set(2, 4, ALIVE)
```

```
columns = ColumnPrinter()
for i in range(5):
    columns.append(str(grid))
    grid = simulate_phased_pipeline(
        grid, in_queue, logic_queue, out_queue)

print(columns)

for thread in threads:
    in_queue.close()
for thread in threads:
    logic_queue.close()
for thread in threads:
    thread.join()
```

```
>>>
    0     |     1     |     2     |     3     |     4
---*----- | --------- | --------- | --------- | ---------
----*---- | --*-*---- | ----*---- | ---*----- | ----*----
--***---- | ---**---- | --*-*---- | ----**--- | -----*---
--------- | ---*----- | ---**---- | ---**---- | ---***---
--------- | --------- | --------- | --------- | ---------
```

Again, this works as expected, but it required a lot of changes and boilerplate. The point here is that Queue does make it possible to solve fan-out and fan-in problems, but the overhead is very high. Although using Queue is a better approach than using Thread instances on their own, it's still not nearly as good as some of the other tools provided by Python (see Item 59: "Consider ThreadPoolExecutor When Threads Are Necessary for Concurrency" and Item 60: "Achieve Highly Concurrent I/O with Coroutines").

Things to Remember

✦ Using Queue instances with a fixed number of worker threads improves the scalability of fan-out and fan-in using threads.

✦ It takes a significant amount of work to refactor existing code to use Queue, especially when multiple stages of a pipeline are required.

✦ Using Queue fundamentally limits the total amount of I/O parallelism a program can leverage compared to alternative approaches provided by other built-in Python features and modules.

Item 59: Consider ThreadPoolExecutor When Threads Are Necessary for Concurrency

Python includes the concurrent.futures built-in module, which provides the ThreadPoolExecutor class. It combines the best of the Thread (see Item 57: "Avoid Creating New Thread Instances for On-demand Fan-out") and Queue (see Item 58: "Understand How Using Queue for Concurrency Requires Refactoring") approaches to solving the parallel I/O problem from the Game of Life example (see Item 56: "Know How to Recognize When Concurrency Is Necessary" for background and the implementations of various functions and classes below):

```
ALIVE = '*'
EMPTY = '-'

class Grid:
    ...

class LockingGrid(Grid):
    ...

def count_neighbors(y, x, get):
    ...

def game_logic(state, neighbors):
    ...
    # Do some blocking input/output in here:
    data = my_socket.recv(100)
    ...

def step_cell(y, x, get, set):
    state = get(y, x)
    neighbors = count_neighbors(y, x, get)
    next_state = game_logic(state, neighbors)
    set(y, x, next_state)
```

Instead of starting a new Thread instance for each Grid square, I can *fan out* by submitting a function to an executor that will be run in a separate thread. Later, I can wait for the result of all tasks in order to fan in:

```
from concurrent.futures import ThreadPoolExecutor

def simulate_pool(pool, grid):
    next_grid = LockingGrid(grid.height, grid.width)
```

```
    futures = []
    for y in range(grid.height):
        for x in range(grid.width):
            args = (y, x, grid.get, next_grid.set)
            future = pool.submit(step_cell, *args)   # Fan out
            futures.append(future)

    for future in futures:
        future.result()                              # Fan in

    return next_grid
```

The threads used for the executor can be allocated in advance, which means I don't have to pay the startup cost on each execution of simulate_pool. I can also specify the maximum number of threads to use for the pool—using the max_workers parameter—to prevent the memory blow-up issues associated with the naive Thread solution to the parallel I/O problem:

```
class ColumnPrinter:
    ...

grid = LockingGrid(5, 9)
grid.set(0, 3, ALIVE)
grid.set(1, 4, ALIVE)
grid.set(2, 2, ALIVE)
grid.set(2, 3, ALIVE)
grid.set(2, 4, ALIVE)

columns = ColumnPrinter()
with ThreadPoolExecutor(max_workers=10) as pool:
    for i in range(5):
        columns.append(str(grid))
        grid = simulate_pool(pool, grid)

print(columns)

>>>
    0      |    1      |    2      |    3      |    4
---*----- | --------- | --------- | --------- | ---------
----*---- | --*-*---- | ----*---- | ---*----- | ----*----
--***---- | ---**---- | --*-*---- | ----**--- | -----*---
--------- | ---*----- | ---**---- | ---**---- | ---***---
--------- | --------- | --------- | --------- | ---------
```

The best part about the ThreadPoolExecutor class is that it automatically propagates exceptions back to the caller when the result method is called on the Future instance returned by the submit method:

```
def game_logic(state, neighbors):
    ...
    raise OSError('Problem with I/O')
    ...

with ThreadPoolExecutor(max_workers=10) as pool:
    task = pool.submit(game_logic, ALIVE, 3)
    task.result()

>>>
Traceback ...
OSError: Problem with I/O
```

If I needed to provide I/O parallelism for the count_neighbors function in addition to game_logic, no modifications to the program would be required since ThreadPoolExecutor already runs these functions concurrently as part of step_cell. It's even possible to achieve CPU parallelism by using the same interface if necessary (see Item 64: "Consider concurrent.futures for True Parallelism").

However, the big problem that remains is the limited amount of I/O parallelism that ThreadPoolExecutor provides. Even if I use a max_workers parameter of 100, this solution still won't scale if I need 10,000+ cells in the grid that require simultaneous I/O. ThreadPoolExecutor is a good choice for situations where there is no asynchronous solution (e.g., file I/O), but there are better ways to maximize I/O parallelism in many cases (see Item 60: "Achieve Highly Concurrent I/O with Coroutines").

Things to Remember

✦ ThreadPoolExecutor enables simple I/O parallelism with limited refactoring, easily avoiding the cost of thread startup each time fanout concurrency is required.

✦ Although ThreadPoolExecutor eliminates the potential memory blow-up issues of using threads directly, it also limits I/O parallelism by requiring max_workers to be specified upfront.

Item 60: Achieve Highly Concurrent I/O with Coroutines

The previous items have tried to solve the parallel I/O problem for the Game of Life example with varying degrees of success. (See Item 56: "Know How to Recognize When Concurrency Is Necessary" for background and the implementations of various functions and classes below.) All of the other approaches fall short in their ability to handle thousands of simultaneously concurrent functions (see Item 57: "Avoid Creating New Thread Instances for On-demand Fan-out," Item 58: "Understand How Using Queue for Concurrency Requires Refactoring," and Item 59: "Consider ThreadPoolExecutor When Threads Are Necessary for Concurrency").

Python addresses the need for highly concurrent I/O with *coroutines*. Coroutines let you have a very large number of seemingly simultaneous functions in your Python programs. They're implemented using the async and await keywords along with the same infrastructure that powers generators (see Item 30: "Consider Generators Instead of Returning Lists," Item 34: "Avoid Injecting Data into Generators with send," and Item 35: "Avoid Causing State Transitions in Generators with throw").

The cost of starting a coroutine is a function call. Once a coroutine is active, it uses less than 1 KB of memory until it's exhausted. Like threads, coroutines are independent functions that can consume inputs from their environment and produce resulting outputs. The difference is that coroutines pause at each await expression and resume executing an async function after the pending *awaitable* is resolved (similar to how yield behaves in generators).

Many separate async functions advanced in lockstep all seem to run simultaneously, mimicking the concurrent behavior of Python threads. However, coroutines do this without the memory overhead, startup and context switching costs, or complex locking and synchronization code that's required for threads. The magical mechanism powering coroutines is the *event loop*, which can do highly concurrent I/O efficiently, while rapidly interleaving execution between appropriately written functions.

I can use coroutines to implement the Game of Life. My goal is to allow for I/O to occur within the game_logic function while overcoming the problems from the Thread and Queue approaches in the previous items. To do this, first I indicate that the game_logic function is a coroutine by defining it using async def instead of def. This will allow me to use the await syntax for I/O, such as an asynchronous read from a socket:

```
ALIVE = '*'
EMPTY = '-'
```

```
class Grid:
    ...

def count_neighbors(y, x, get):
    ...

async def game_logic(state, neighbors):
    ...
    # Do some input/output in here:
    data = await my_socket.read(50)
    ...
```

Similarly, I can turn step_cell into a coroutine by adding async to its definition and using await for the call to the game_logic function:

```
async def step_cell(y, x, get, set):
    state = get(y, x)
    neighbors = count_neighbors(y, x, get)
    next_state = await game_logic(state, neighbors)
    set(y, x, next_state)
```

The simulate function also needs to become a coroutine:

```
import asyncio

async def simulate(grid):
    next_grid = Grid(grid.height, grid.width)

    tasks = []
    for y in range(grid.height):
        for x in range(grid.width):
            task = step_cell(
                y, x, grid.get, next_grid.set)       # Fan out
            tasks.append(task)

    await asyncio.gather(*tasks)                      # Fan in

    return next_grid
```

The coroutine version of the simulate function requires some explanation:

- Calling step_cell doesn't immediately run that function. Instead, it returns a coroutine instance that can be used with an await expression at a later time. This is similar to how generator functions that use yield return a generator instance when they're called instead of executing immediately. Deferring execution like this is the mechanism that causes *fan-out*.

- The gather function from the asyncio built-in library causes *fan-in*. The await expression on gather instructs the event loop to run the step_cell coroutines concurrently and resume execution of the simulate coroutine when all of them have been completed.

- No locks are required for the Grid instance since all execution occurs within a single thread. The I/O becomes parallelized as part of the event loop that's provided by asyncio.

Finally, I can drive this code with a one-line change to the original example. This relies on the asyncio.run function to execute the simulate coroutine in an event loop and carry out its dependent I/O:

```
class ColumnPrinter:
    ...

grid = Grid(5, 9)
grid.set(0, 3, ALIVE)
grid.set(1, 4, ALIVE)
grid.set(2, 2, ALIVE)
grid.set(2, 3, ALIVE)
grid.set(2, 4, ALIVE)

columns = ColumnPrinter()
for i in range(5):
    columns.append(str(grid))
    grid = asyncio.run(simulate(grid))   # Run the event loop

print(columns)
```

```
>>>
    0     |    1     |    2     |    3     |    4
---*----- | --------- | --------- | --------- | ---------
----*---- | --*-*---- | ----*---- | ---*----- | ----*----
--***---- | ---**---- | --*-*---- | ----**--- | -----*---
--------- | ---*----- | ---**---- | ---**---- | ---***---
--------- | --------- | --------- | --------- | ---------
```

The result is the same as before. All of the overhead associated with threads has been eliminated. Whereas the Queue and ThreadPoolExecutor approaches are limited in their exception handling—merely re-raising exceptions across thread boundaries—with coroutines I can actually use the interactive debugger to step through the code line by line (see Item 80: "Consider Interactive Debugging with pdb"):

```
async def game_logic(state, neighbors):
    ...
```

```
    raise OSError('Problem with I/O')
    ...
```

```
asyncio.run(game_logic(ALIVE, 3))
```

```
>>>
Traceback ...
OSError: Problem with I/O
```

Later, if my requirements change and I also need to do I/O from within count_neighbors, I can easily accomplish this by adding async and await keywords to the existing functions and call sites instead of having to restructure everything as I would have had to do if I were using Thread or Queue instances (see Item 61: "Know How to Port Threaded I/O to asyncio" for another example):

```
async def count_neighbors(y, x, get):
    ...

async def step_cell(y, x, get, set):
    state = get(y, x)
    neighbors = await count_neighbors(y, x, get)
    next_state = await game_logic(state, neighbors)
    set(y, x, next_state)

grid = Grid(5, 9)
grid.set(0, 3, ALIVE)
grid.set(1, 4, ALIVE)
grid.set(2, 2, ALIVE)
grid.set(2, 3, ALIVE)
grid.set(2, 4, ALIVE)

columns = ColumnPrinter()
for i in range(5):
    columns.append(str(grid))
    grid = asyncio.run(simulate(grid))

print(columns)
```

```
>>>
    0     |    1     |    2     |    3     |    4
---*----- | --------- | --------- | --------- | ---------
----*---- | --*-*---- | ----*---- | ---*----- | ----*----
--***---- | ---**---- | --*-*---- | ----**--- | -----*---
--------- | ---*----- | ---**---- | ---**---- | ---***---
--------- | --------- | --------- | --------- | ---------
```

The beauty of coroutines is that they decouple your code's instructions for the external environment (i.e., I/O) from the implementation that carries out your wishes (i.e., the event loop). They let you focus on the logic of what you're trying to do instead of wasting time trying to figure out how you're going to accomplish your goals concurrently.

Things to Remember

✦ Functions that are defined using the async keyword are called coroutines. A caller can receive the result of a dependent coroutine by using the await keyword.

✦ Coroutines provide an efficient way to run tens of thousands of functions seemingly at the same time.

✦ Coroutines can use fan-out and fan-in in order to parallelize I/O, while also overcoming all of the problems associated with doing I/O in threads.

Item 61: Know How to Port Threaded I/O to asyncio

Once you understand the advantage of coroutines (see Item 60: "Achieve Highly Concurrent I/O with Coroutines"), it may seem daunting to port an existing codebase to use them. Luckily, Python's support for asynchronous execution is well integrated into the language. This makes it straightforward to move code that does threaded, blocking I/O over to coroutines and asynchronous I/O.

For example, say that I have a TCP-based server for playing a game involving guessing a number. The server takes lower and upper parameters that determine the range of numbers to consider. Then, the server returns guesses for integer values in that range as they are requested by the client. Finally, the server collects reports from the client on whether each of those numbers was closer (warmer) or further away (colder) from the client's secret number.

The most common way to build this type of client/server system is by using blocking I/O and threads (see Item 53: "Use Threads for Blocking I/O, Avoid for Parallelism"). To do this, I need a helper class that can manage sending and receiving of messages. For my purposes, each line sent or received represents a command to be processed:

```
class EOFError(Exception):
    pass

class ConnectionBase:
    def __init__(self, connection):
```

```
        self.connection = connection
        self.file = connection.makefile('rb')

    def send(self, command):
        line = command + '\n'
        data = line.encode()
        self.connection.send(data)

    def receive(self):
        line = self.file.readline()
        if not line:
            raise EOFError('Connection closed')
        return line[:-1].decode()
```

The server is implemented as a class that handles one connection at a time and maintains the client's session state:

```
import random

WARMER = 'Warmer'
COLDER = 'Colder'
UNSURE = 'Unsure'
CORRECT = 'Correct'

class UnknownCommandError(Exception):
    pass

class Session(ConnectionBase):
    def __init__(self, *args):
        super().__init__(*args)
        self._clear_state(None, None)

    def _clear_state(self, lower, upper):
        self.lower = lower
        self.upper = upper
        self.secret = None
        self.guesses = []
```

It has one primary method that handles incoming commands from the client and dispatches them to methods as needed. Note that here I'm using an assignment expression (introduced in Python 3.8; see Item 10: "Prevent Repetition with Assignment Expressions") to keep the code short:

```
    def loop(self):
        while command := self.receive():
```

```
    parts = command.split(' ')
    if parts[0] == 'PARAMS':
        self.set_params(parts)
    elif parts[0] == 'NUMBER':
        self.send_number()
    elif parts[0] == 'REPORT':
        self.receive_report(parts)
    else:
        raise UnknownCommandError(command)
```

The first command sets the lower and upper bounds for the numbers that the server is trying to guess:

```
def set_params(self, parts):
    assert len(parts) == 3
    lower = int(parts[1])
    upper = int(parts[2])
    self._clear_state(lower, upper)
```

The second command makes a new guess based on the previous state that's stored in the client's Session instance. Specifically, this code ensures that the server will never try to guess the same number more than once per parameter assignment:

```
def next_guess(self):
    if self.secret is not None:
        return self.secret

    while True:
        guess = random.randint(self.lower, self.upper)
        if guess not in self.guesses:
            return guess

def send_number(self):
    guess = self.next_guess()
    self.guesses.append(guess)
    self.send(format(guess))
```

The third command receives the decision from the client of whether the guess was warmer or colder, and it updates the Session state accordingly:

```
def receive_report(self, parts):
    assert len(parts) == 2
    decision = parts[1]

    last = self.guesses[-1]
```

```
    if decision == CORRECT:
        self.secret = last

    print(f'Server: {last} is {decision}')
```

The client is also implemented using a stateful class:

```
import contextlib
import math

class Client(ConnectionBase):
    def __init__(self, *args):
        super().__init__(*args)
        self._clear_state()

    def _clear_state(self):
        self.secret = None
        self.last_distance = None
```

The parameters of each guessing game are set using a with statement to ensure that state is correctly managed on the server side (see Item 66: "Consider contextlib and with Statements for Reusable try/finally Behavior" for background and Item 63: "Avoid Blocking the asyncio Event Loop to Maximize Responsiveness" for another example). This method sends the first command to the server:

```
    @contextlib.contextmanager
    def session(self, lower, upper, secret):
        print(f'Guess a number between {lower} and {upper}!'
              f' Shhhhh, it\'s {secret}.')
        self.secret = secret
        self.send(f'PARAMS {lower} {upper}')
        try:
            yield
        finally:
            self._clear_state()
            self.send('PARAMS 0 -1')
```

New guesses are requested from the server, using another method that implements the second command:

```
    def request_numbers(self, count):
        for _ in range(count):
            self.send('NUMBER')
            data = self.receive()
            yield int(data)
            if self.last_distance == 0:
                return
```

Whether each guess from the server was warmer or colder than the last is reported using the third command in the final method:

```
def report_outcome(self, number):
    new_distance = math.fabs(number - self.secret)
    decision = UNSURE

    if new_distance == 0:
        decision = CORRECT
    elif self.last_distance is None:
        pass
    elif new_distance < self.last_distance:
        decision = WARMER
    elif new_distance > self.last_distance:
        decision = COLDER

    self.last_distance = new_distance

    self.send(f'REPORT {decision}')
    return decision
```

I can run the server by having one thread listen on a socket and spawn additional threads to handle the new connections:

```
import socket
from threading import Thread

def handle_connection(connection):
    with connection:
        session = Session(connection)
        try:
            session.loop()
        except EOFError:
            pass

def run_server(address):
    with socket.socket() as listener:
        listener.bind(address)
        listener.listen()
        while True:
            connection, _ = listener.accept()
            thread = Thread(target=handle_connection,
                            args=(connection,),
                            daemon=True)
            thread.start()
```

The client runs in the main thread and returns the results of the guessing game to the caller. This code explicitly exercises a variety of Python language features (for loops, with statements, generators, comprehensions) so that below I can show what it takes to port these over to using coroutines:

```python
def run_client(address):
    with socket.create_connection(address) as connection:
        client = Client(connection)

        with client.session(1, 5, 3):
            results = [(x, client.report_outcome(x))
                       for x in client.request_numbers(5)]

        with client.session(10, 15, 12):
            for number in client.request_numbers(5):
                outcome = client.report_outcome(number)
                results.append((number, outcome))

    return results
```

Finally, I can glue all of this together and confirm that it works as expected:

```python
def main():
    address = ('127.0.0.1', 1234)
    server_thread = Thread(
        target=run_server, args=(address,), daemon=True)
    server_thread.start()

    results = run_client(address)
    for number, outcome in results:
        print(f'Client: {number} is {outcome}')

main()

>>>
Guess a number between 1 and 5! Shhhhh, it's 3.
Server: 4 is Unsure
Server: 1 is Colder
Server: 5 is Unsure
Server: 3 is Correct
Guess a number between 10 and 15! Shhhhh, it's 12.
Server: 11 is Unsure
Server: 10 is Colder
Server: 12 is Correct
```

```
Client: 4 is Unsure
Client: 1 is Colder
Client: 5 is Unsure
Client: 3 is Correct
Client: 11 is Unsure
Client: 10 is Colder
Client: 12 is Correct
```

How much effort is needed to convert this example to using async, await, and the asyncio built-in module?

First, I need to update my ConnectionBase class to provide coroutines for send and receive instead of blocking I/O methods. I've marked each line that's changed with a # Changed comment to make it clear what the delta is between this new example and the code above:

```
class AsyncConnectionBase:
    def __init__(self, reader, writer):            # Changed
        self.reader = reader                       # Changed
        self.writer = writer                       # Changed

    async def send(self, command):
        line = command + '\n'
        data = line.encode()
        self.writer.write(data)                    # Changed
        await self.writer.drain()                  # Changed

    async def receive(self):
        line = await self.reader.readline()        # Changed
        if not line:
            raise EOFError('Connection closed')
        return line[:-1].decode()
```

I can create another stateful class to represent the session state for a single connection. The only changes here are the class's name and inheriting from AsyncConnectionBase instead of ConnectionBase:

```
class AsyncSession(AsyncConnectionBase):           # Changed
    def __init__(self, *args):
        ...

    def _clear_values(self, lower, upper):
        ...
```

The primary entry point for the server's command processing loop requires only minimal changes to become a coroutine:

```
    async def loop(self):                          # Changed
```

```
    while command := await self.receive():        # Changed
        parts = command.split(' ')
        if parts[0] == 'PARAMS':
            self.set_params(parts)
        elif parts[0] == 'NUMBER':
            await self.send_number()                # Changed
        elif parts[0] == 'REPORT':
            self.receive_report(parts)
        else:
            raise UnknownCommandError(command)
```

No changes are required for handling the first command:

```
def set_params(self, parts):
    ...
```

The only change required for the second command is allowing asynchronous I/O to be used when guesses are transmitted to the client:

```
def next_guess(self):
    ...


async def send_number(self):                       # Changed
    guess = self.next_guess()
    self.guesses.append(guess)
    await self.send(format(guess))                 # Changed
```

No changes are required for processing the third command:

```
def receive_report(self, parts):
    ...
```

Similarly, the client class needs to be reimplemented to inherit from AsyncConnectionBase:

```
class AsyncClient(AsyncConnectionBase):            # Changed
    def __init__(self, *args):
        ...


    def _clear_state(self):
        ...
```

The first command method for the client requires a few async and await keywords to be added. It also needs to use the asynccontextmanager helper function from the contextlib built-in module:

```
@contextlib.asynccontextmanager                    # Changed
async def session(self, lower, upper, secret):  # Changed
    print(f'Guess a number between {lower} and {upper}!'
          f' Shhhhh, it\'s {secret}.')
```

```
        self.secret = secret
        await self.send(f'PARAMS {lower} {upper}')   # Changed
        try:
            yield
        finally:
            self._clear_state()
            await self.send('PARAMS 0 -1')            # Changed
```

The second command again only requires the addition of async and await anywhere coroutine behavior is required:

```
    async def request_numbers(self, count):          # Changed
        for _ in range(count):
            await self.send('NUMBER')                 # Changed
            data = await self.receive()               # Changed
            yield int(data)
            if self.last_distance == 0:
                return
```

The third command only requires adding one async and one await keyword:

```
    async def report_outcome(self, number):          # Changed
        ...
        await self.send(f'REPORT {decision}')         # Changed
        ...
```

The code that runs the server needs to be completely reimplemented to use the asyncio built-in module and its start_server function:

```
import asyncio

async def handle_async_connection(reader, writer):
    session = AsyncSession(reader, writer)
    try:
        await session.loop()
    except EOFError:
        pass

async def run_async_server(address):
    server = await asyncio.start_server(
        handle_async_connection, *address)
    async with server:
        await server.serve_forever()
```

The run_client function that initiates the game requires changes on nearly every line. Any code that previously interacted with the blocking socket instances has to be replaced with asyncio versions of similar functionality (which are marked with # New below). All other lines in the function that require interaction with coroutines need to use async and await keywords as appropriate. If you forget to add one of these keywords in a necessary place, an exception will be raised at runtime.

```
async def run_async_client(address):
    streams = await asyncio.open_connection(*address)    # New
    client = AsyncClient(*streams)                        # New

    async with client.session(1, 5, 3):
        results = [(x, await client.report_outcome(x))
                      async for x in client.request_numbers(5)]

    async with client.session(10, 15, 12):
        async for number in client.request_numbers(5):
            outcome = await client.report_outcome(number)
            results.append((number, outcome))

    _, writer = streams                                   # New
    writer.close()                                        # New
    await writer.wait_closed()                            # New

    return results
```

What's most interesting about run_async_client is that I didn't have to restructure any of the substantive parts of interacting with the AsyncClient in order to port this function over to use coroutines. Each of the language features that I needed has a corresponding asynchronous version, which made the migration easy to do.

This won't always be the case, though. There are currently no asynchronous versions of the next and iter built-in functions (see Item 31: "Be Defensive When Iterating Over Arguments" for background); you have to await on the __anext__ and __aiter__ methods directly. There's also no asynchronous version of yield from (see Item 33: "Compose Multiple Generators with yield from"), which makes it noisier to compose generators. But given the rapid pace at which async functionality is being added to Python, it's only a matter of time before these features become available.

Finally, the glue needs to be updated to run this new asynchronous example end-to-end. I use the asyncio.create_task function to enqueue the server for execution on the event loop so that it runs in parallel with the client when the await expression is reached. This is another approach to causing fan-out with different behavior than the asyncio.gather function:

```
async def main_async():
    address = ('127.0.0.1', 4321)

    server = run_async_server(address)
    asyncio.create_task(server)

    results = await run_async_client(address)
    for number, outcome in results:
        print(f'Client: {number} is {outcome}')

asyncio.run(main_async())

>>>
Guess a number between 1 and 5! Shhhhh, it's 3.
Server: 5 is Unsure
Server: 4 is Warmer
Server: 2 is Unsure
Server: 1 is Colder
Server: 3 is Correct
Guess a number between 10 and 15! Shhhhh, it's 12.
Server: 14 is Unsure
Server: 10 is Unsure
Server: 15 is Colder
Server: 12 is Correct
Client: 5 is Unsure
Client: 4 is Warmer
Client: 2 is Unsure
Client: 1 is Colder
Client: 3 is Correct
Client: 14 is Unsure
Client: 10 is Unsure
Client: 15 is Colder
Client: 12 is Correct
```

This works as expected. The coroutine version is easier to follow because all of the interactions with threads have been removed. The asyncio built-in module also provides many helper functions and shortens the amount of socket boilerplate required to write a server like this.

Your use case may be more complex and harder to port for a variety of reasons. The asyncio module has a vast number of I/O, synchronization, and task management features that could make adopting coroutines easier for you (see Item 62: "Mix Threads and Coroutines to Ease the Transition to asyncio" and Item 63: "Avoid Blocking the asyncio Event Loop to Maximize Responsiveness"). Be sure to check out the online documentation for the library to understand its full potential.

Things to Remember

✦ Python provides asynchronous versions of for loops, with statements, generators, comprehensions, and library helper functions that can be used as drop-in replacements in coroutines.

✦ The asyncio built-in module makes it straightforward to port existing code that uses threads and blocking I/O over to coroutines and asynchronous I/O.

Item 62: Mix Threads and Coroutines to Ease the Transition to asyncio

In the previous item (see Item 61: "Know How to Port Threaded I/O to asyncio"), I ported a TCP server that does blocking I/O with threads over to use asyncio with coroutines. The transition was big-bang: I moved all of the code to the new style in one go. But it's rarely feasible to port a large program this way. Instead, you usually need to incrementally migrate your codebase while also updating your tests as needed and verifying that everything works at each step along the way.

In order to do that, your codebase needs to be able to use threads for blocking I/O (see Item 53: "Use Threads for Blocking I/O, Avoid for Parallelism") and coroutines for asynchronous I/O (see Item 60: "Achieve Highly Concurrent I/O with Coroutines") at the same time in a way that's mutually compatible. Practically, this means that you need threads to be able to run coroutines, and you need coroutines to be able to start and wait on threads. Luckily, asyncio includes built-in facilities for making this type of interoperability straightforward.

For example, say that I'm writing a program that merges log files into one output stream to aid with debugging. Given a file handle for an input log, I need a way to detect whether new data is available and return the next line of input. I can do this using the tell method of the file handle to check whether the current read position matches the length of the file. When no new data is present, an exception should be raised (see Item 20: "Prefer Raising Exceptions to Returning None" for background):

```
class NoNewData(Exception):
    pass
```

```
def readline(handle):
    offset = handle.tell()
    handle.seek(0, 2)
    length = handle.tell()

    if length == offset:
        raise NoNewData

    handle.seek(offset, 0)
    return handle.readline()
```

By wrapping this function in a while loop, I can turn it into a worker thread. When a new line is available, I call a given callback function to write it to the output log (see Item 38: "Accept Functions Instead of Classes for Simple Interfaces" for why to use a function interface for this instead of a class). When no data is available, the thread sleeps to reduce the amount of busy waiting caused by polling for new data. When the input file handle is closed, the worker thread exits:

```
import time

def tail_file(handle, interval, write_func):
    while not handle.closed:
        try:
            line = readline(handle)
        except NoNewData:
            time.sleep(interval)
        else:
            write_func(line)
```

Now, I can start one worker thread per input file and unify their output into a single output file. The write helper function below needs to use a Lock instance (see Item 54: "Use Lock to Prevent Data Races in Threads") in order to serialize writes to the output stream and make sure that there are no intra-line conflicts:

```
from threading import Lock, Thread

def run_threads(handles, interval, output_path):
    with open(output_path, 'wb') as output:
        lock = Lock()
        def write(data):
            with lock:
                output.write(data)
```

```
    threads = []
    for handle in handles:
        args = (handle, interval, write)
        thread = Thread(target=tail_file, args=args)
        thread.start()
        threads.append(thread)

    for thread in threads:
        thread.join()
```

As long as an input file handle is still alive, its corresponding worker thread will also stay alive. That means it's sufficient to wait for the join method from each thread to complete in order to know that the whole process is done.

Given a set of input paths and an output path, I can call `run_threads` and confirm that it works as expected. How the input file handles are created or separately closed isn't important in order to demonstrate this code's behavior, nor is the output verification function—defined in `confirm_merge` that follows—which is why I've left them out here:

```
def confirm_merge(input_paths, output_path):
    ...

input_paths = ...
handles = ...
output_path = ...
run_threads(handles, 0.1, output_path)

confirm_merge(input_paths, output_path)
```

With this threaded implementation as the starting point, how can I incrementally convert this code to use asyncio and coroutines instead? There are two approaches: top-down and bottom-up.

Top-down means starting at the highest parts of a codebase, like in the main entry points, and working down to the individual functions and classes that are the leaves of the call hierarchy. This approach can be useful when you maintain a lot of common modules that you use across many different programs. By porting the entry points first, you can wait to port the common modules until you're already using coroutines everywhere else.

The concrete steps are:

1. Change a top function to use `async def` instead of `def`.

2. Wrap all of its calls that do I/O—potentially blocking the event loop—to use `asyncio.run_in_executor` instead.

3. Ensure that the resources or callbacks used by run_in_executor invocations are properly synchronized (i.e., using Lock or the asyncio.run_coroutine_threadsafe function).

4. Try to eliminate get_event_loop and run_in_executor calls by moving downward through the call hierarchy and converting intermediate functions and methods to coroutines (following the first three steps).

Here, I apply steps 1–3 to the run_threads function:

```
import asyncio

async def run_tasks_mixed(handles, interval, output_path):
    loop = asyncio.get_event_loop()

    with open(output_path, 'wb') as output:
        async def write_async(data):
            output.write(data)

        def write(data):
            coro = write_async(data)
            future = asyncio.run_coroutine_threadsafe(
                coro, loop)
            future.result()

        tasks = []
        for handle in handles:
            task = loop.run_in_executor(
                None, tail_file, handle, interval, write)
            tasks.append(task)

        await asyncio.gather(*tasks)
```

The run_in_executor method instructs the event loop to run a given function—tail_file in this case—using a specific ThreadPoolExecutor (see Item 59: "Consider ThreadPoolExecutor When Threads Are Necessary for Concurrency") or a default executor instance when the first parameter is None. By making multiple calls to run_in_executor without corresponding await expressions, the run_tasks_mixed coroutine fans out to have one concurrent line of execution for each input file. Then, the asyncio.gather function along with an await expression fans in the tail_file threads until they all complete (see Item 56: "Know How to Recognize When Concurrency Is Necessary" for more about fan-out and fan-in).

This code eliminates the need for the Lock instance in the write helper by using asyncio.run_coroutine_threadsafe. This function allows plain old worker threads to call a coroutine—write_async in this case—and have it execute in the event loop from the main thread (or from any other thread, if necessary). This effectively synchronizes the threads together and ensures that all writes to the output file are only done by the event loop in the main thread. Once the asyncio.gather awaitable is resolved, I can assume that all writes to the output file have also completed, and thus I can close the output file handle in the with statement without having to worry about race conditions.

I can verify that this code works as expected. I use the asyncio.run function to start the coroutine and run the main event loop:

```
input_paths = ...
handles = ...
output_path = ...
asyncio.run(run_tasks_mixed(handles, 0.1, output_path))

confirm_merge(input_paths, output_path)
```

Now, I can apply step 4 to the run_tasks_mixed function by moving down the call stack. I can redefine the tail_file dependent function to be an asynchronous coroutine instead of doing blocking I/O by following steps 1–3:

```
async def tail_async(handle, interval, write_func):
    loop = asyncio.get_event_loop()

    while not handle.closed:
        try:
            line = await loop.run_in_executor(
                None, readline, handle)
        except NoNewData:
            await asyncio.sleep(interval)
        else:
            await write_func(line)
```

This new implementation of tail_async allows me to push calls to get_event_loop and run_in_executor down the stack and out of the run_tasks_mixed function entirely. What's left is clean and much easier to follow:

```
async def run_tasks(handles, interval, output_path):
    with open(output_path, 'wb') as output:
        async def write_async(data):
            output.write(data)
```

```
        tasks = []
        for handle in handles:
            coro = tail_async(handle, interval, write_async)
            task = asyncio.create_task(coro)
            tasks.append(task)

        await asyncio.gather(*tasks)
```

I can verify that run_tasks works as expected, too:

```
input_paths = ...
handles = ...
output_path = ...
asyncio.run(run_tasks(handles, 0.1, output_path))

confirm_merge(input_paths, output_path)
```

It's possible to continue this iterative refactoring pattern and convert readline into an asynchronous coroutine as well. However, that function requires so many blocking file I/O operations that it doesn't seem worth porting, given how much that would reduce the clarity of the code and hurt performance. In some situations, it makes sense to move everything to asyncio, and in others it doesn't.

The bottom-up approach to adopting coroutines has four steps that are similar to the steps of the top-down style, but the process traverses the call hierarchy in the opposite direction: from leaves to entry points.

The concrete steps are:

1. Create a new asynchronous coroutine version of each leaf function that you're trying to port.

2. Change the existing synchronous functions so they call the coroutine versions and run the event loop instead of implementing any real behavior.

3. Move up a level of the call hierarchy, make another layer of coroutines, and replace existing calls to synchronous functions with calls to the coroutines defined in step 1.

4. Delete synchronous wrappers around coroutines created in step 2 as you stop requiring them to glue the pieces together.

For the example above, I would start with the tail_file function since I decided that the readline function should keep using blocking I/O. I can rewrite tail_file so it merely wraps the tail_async coroutine that I defined above. To run that coroutine until it finishes, I need to create an event loop for each `tail_file` worker thread and then call its `run_until_complete` method. This method will block the current thread and drive the event loop until the `tail_async` coroutine exits, achieving the same behavior as the threaded, blocking I/O version of `tail_file`:

```
def tail_file(handle, interval, write_func):
    loop = asyncio.new_event_loop()
    asyncio.set_event_loop(loop)

    async def write_async(data):
        write_func(data)

    coro = tail_async(handle, interval, write_async)
    loop.run_until_complete(coro)
```

This new `tail_file` function is a drop-in replacement for the old one. I can verify that everything works as expected by calling `run_threads` again:

```
input_paths = ...
handles = ...
output_path = ...
run_threads(handles, 0.1, output_path)

confirm_merge(input_paths, output_path)
```

After wrapping `tail_async` with `tail_file`, the next step is to convert the `run_threads` function to a coroutine. This ends up being the same work as step 4 of the top-down approach above, so at this point, the styles converge.

This is a great start for adopting asyncio, but there's even more that you could do to increase the responsiveness of your program (see Item 63: "Avoid Blocking the asyncio Event Loop to Maximize Responsiveness").

Things to Remember

✦ The awaitable `run_in_executor` method of the asyncio event loop enables coroutines to run synchronous functions in `ThreadPoolExecutor` pools. This facilitates top-down migrations to asyncio.

✦ The `run_until_complete` method of the asyncio event loop enables synchronous code to run a coroutine until it finishes. The `asyncio.run_coroutine_threadsafe` function provides the same functionality across thread boundaries. Together these help with bottom-up migrations to asyncio.

Item 63: Avoid Blocking the asyncio Event Loop to Maximize Responsiveness

In the previous item I showed how to migrate to asyncio incrementally (see Item 62: "Mix Threads and Coroutines to Ease the Transition to asyncio" for background and the implementation of various functions below). The resulting coroutine properly tails input files and merges them into a single output:

```
import asyncio

async def run_tasks(handles, interval, output_path):
    with open(output_path, 'wb') as output:
        async def write_async(data):
            output.write(data)

        tasks = []
        for handle in handles:
            coro = tail_async(handle, interval, write_async)
            task = asyncio.create_task(coro)
            tasks.append(task)

        await asyncio.gather(*tasks)
```

However, it still has one big problem: The open, close, and write calls for the output file handle happen in the main event loop. These operations all require making system calls to the program's host operating system, which may block the event loop for significant amounts of time and prevent other coroutines from making progress. This could hurt overall responsiveness and increase latency, especially for programs such as highly concurrent servers.

I can detect when this problem happens by passing the debug=True parameter to the asyncio.run function. Here, I show how the file and line of a bad coroutine, presumably blocked on a slow system call, can be identified:

```
import time

async def slow_coroutine():
    time.sleep(0.5)  # Simulating slow I/O

asyncio.run(slow_coroutine(), debug=True)

>>>
Executing <Task finished name='Task-1' coro=<slow_coroutine()
➡done, defined at example.py:29> result=None created
➡at .../asyncio/base_events.py:487> took 0.503 seconds
...
```

If I want the most responsive program possible, I need to minimize the potential system calls that are made from within the event loop. In this case, I can create a new Thread subclass (see Item 53: "Use Threads for Blocking I/O, Avoid for Parallelism") that encapsulates everything required to write to the output file using its own event loop:

```
from threading import Thread

class WriteThread(Thread):
    def __init__(self, output_path):
        super().__init__()
        self.output_path = output_path
        self.output = None
        self.loop = asyncio.new_event_loop()

    def run(self):
        asyncio.set_event_loop(self.loop)
        with open(self.output_path, 'wb') as self.output:
            self.loop.run_forever()

        # Run one final round of callbacks so the await on
        # stop() in another event loop will be resolved.
        self.loop.run_until_complete(asyncio.sleep(0))
```

Coroutines in other threads can directly call and await on the write method of this class, since it's merely a thread-safe wrapper around the real_write method that actually does the I/O. This eliminates the need for a Lock (see Item 54: "Use Lock to Prevent Data Races in Threads"):

```
    async def real_write(self, data):
        self.output.write(data)

    async def write(self, data):
        coro = self.real_write(data)
        future = asyncio.run_coroutine_threadsafe(
            coro, self.loop)
        await asyncio.wrap_future(future)
```

Other coroutines can tell the worker thread when to stop in a thread-safe manner, using similar boilerplate:

```
    async def real_stop(self):
        self.loop.stop()
```

```
async def stop(self):
    coro = self.real_stop()
    future = asyncio.run_coroutine_threadsafe(
        coro, self.loop)
    await asyncio.wrap_future(future)
```

I can also define the __aenter__ and __aexit__ methods to allow this class to be used in with statements (see Item 66: "Consider contextlib and with Statements for Reusable try/finally Behavior"). This ensures that the worker thread starts and stops at the right times without slowing down the main event loop thread:

```
async def __aenter__(self):
    loop = asyncio.get_event_loop()
    await loop.run_in_executor(None, self.start)
    return self

async def __aexit__(self, *_):
    await self.stop()
```

With this new WriteThread class, I can refactor run_tasks into a fully asynchronous version that's easy to read and completely avoids running slow system calls in the main event loop thread:

```
def readline(handle):
    ...

async def tail_async(handle, interval, write_func):
    ...

async def run_fully_async(handles, interval, output_path):
    async with WriteThread(output_path) as output:
        tasks = []
        for handle in handles:
            coro = tail_async(handle, interval, output.write)
            task = asyncio.create_task(coro)
            tasks.append(task)

        await asyncio.gather(*tasks)
```

I can verify that this works as expected, given a set of input handles and an output file path:

```
def confirm_merge(input_paths, output_path):
    ...
```

```
input_paths = ...
handles = ...
output_path = ...
asyncio.run(run_fully_async(handles, 0.1, output_path))

confirm_merge(input_paths, output_path)
```

Things to Remember

✦ Making system calls in coroutines—including blocking I/O and starting threads—can reduce program responsiveness and increase the perception of latency.

✦ Pass the debug=True parameter to asyncio.run in order to detect when certain coroutines are preventing the event loop from reacting quickly.

Item 64: Consider concurrent.futures for True Parallelism

At some point in writing Python programs, you may hit the performance wall. Even after optimizing your code (see Item 70: "Profile Before Optimizing"), your program's execution may still be too slow for your needs. On modern computers that have an increasing number of CPU cores, it's reasonable to assume that one solution would be parallelism. What if you could split your code's computation into independent pieces of work that run simultaneously across multiple CPU cores?

Unfortunately, Python's global interpreter lock (GIL) prevents true parallelism in threads (see Item 53: "Use Threads for Blocking I/O, Avoid for Parallelism"), so that option is out. Another common suggestion is to rewrite your most performance-critical code as an extension module, using the C language. C gets you closer to the bare metal and can run faster than Python, eliminating the need for parallelism in some cases. C extensions can also start native threads independent of the Python interpreter that run in parallel and utilize multiple CPU cores with no concern for the GIL. Python's API for C extensions is well documented and a good choice for an escape hatch. It's also worth checking out tools like SWIG and CLIF to aid in extension development.

But rewriting your code in C has a high cost. Code that is short and understandable in Python can become verbose and complicated in C. Such a port requires extensive testing to ensure that the functionality is equivalent to the original Python code and that no bugs have been introduced. Sometimes it's worth it, which explains the large ecosystem of C-extension modules in the Python community that speed up things like text parsing, image compositing, and matrix math. There are even open source tools such as Cython and Numba that can ease the transition to C.

The problem is that moving one piece of your program to C isn't sufficient most of the time. Optimized Python programs usually don't have one major source of slowness; rather, there are often many significant contributors. To get the benefits of C's bare metal and threads, you'd need to port large parts of your program, drastically increasing testing needs and risk. There must be a better way to preserve your investment in Python to solve difficult computational problems.

The multiprocessing built-in module, which is easily accessed via the concurrent.futures built-in module, may be exactly what you need (see Item 59: "Consider ThreadPoolExecutor When Threads Are Necessary for Concurrency" for a related example). It enables Python to utilize multiple CPU cores in parallel by running additional interpreters as child processes. These child processes are separate from the main interpreter, so their global interpreter locks are also separate. Each child can fully utilize one CPU core. Each child has a link to the main process where it receives instructions to do computation and returns results.

For example, say that I want to do something computationally intensive with Python and utilize multiple CPU cores. I'll use an implementation of finding the greatest common divisor of two numbers as a proxy for a more computationally intense algorithm (like simulating fluid dynamics with the Navier–Stokes equation):

```
# my_module.py
def gcd(pair):
    a, b = pair
    low = min(a, b)
    for i in range(low, 0, -1):
        if a % i == 0 and b % i == 0:
            return i
    assert False, 'Not reachable'
```

Running this function in serial takes a linearly increasing amount of time because there is no parallelism:

```
# run_serial.py
import my_module
import time
```

```
NUMBERS = [
    (1963309, 2265973), (2030677, 3814172),
    (1551645, 2229620), (2039045, 2020802),
    (1823712, 1924928), (2293129, 1020491),
    (1281238, 2273782), (3823812, 4237281),
    (3812741, 4729139), (1292391, 2123811),
]

def main():
    start = time.time()
    results = list(map(my_module.gcd, NUMBERS))
    end = time.time()
    delta = end - start
    print(f'Took {delta:.3f} seconds')

if __name__ == '__main__':
    main()

>>>
Took 1.173 seconds
```

Running this code on multiple Python threads will yield no speed improvement because the GIL prevents Python from using multiple CPU cores in parallel. Here, I do the same computation as above but using the concurrent.futures module with its ThreadPoolExecutor class and two worker threads (to match the number of CPU cores on my computer):

```
# run_threads.py
import my_module
from concurrent.futures import ThreadPoolExecutor
import time

NUMBERS = [
    ...
]

def main():
    start = time.time()
    pool = ThreadPoolExecutor(max_workers=2)
    results = list(pool.map(my_module.gcd, NUMBERS))
    end = time.time()
    delta = end - start
    print(f'Took {delta:.3f} seconds')
```

```
if __name__ == '__main__':
    main()
```

```
>>>
Took 1.436 seconds
```

It's even slower this time because of the overhead of starting and communicating with the pool of threads.

Now for the surprising part: Changing a single line of code causes something magical to happen. If I replace the ThreadPoolExecutor with the ProcessPoolExecutor from the concurrent.futures module, everything speeds up:

```
# run_parallel.py
import my_module
from concurrent.futures import ProcessPoolExecutor
import time

NUMBERS = [
    ...
]

def main():
    start = time.time()
    pool = ProcessPoolExecutor(max_workers=2)  # The one change
    results = list(pool.map(my_module.gcd, NUMBERS))
    end = time.time()
    delta = end - start
    print(f'Took {delta:.3f} seconds')

if __name__ == '__main__':
    main()
```

```
>>>
Took 0.683 seconds
```

Running on my dual-core machine, this is significantly faster! How is this possible? Here's what the ProcessPoolExecutor class actually does (via the low-level constructs provided by the multiprocessing module):

1. It takes each item from the numbers input data to map.

2. It serializes the item into binary data by using the pickle module (see Item 68: "Make pickle Reliable with copyreg").

3. It copies the serialized data from the main interpreter process to a child interpreter process over a local socket.

4. It deserializes the data back into Python objects, using `pickle` in the child process.

5. It imports the Python module containing the `gcd` function.

6. It runs the function on the input data in parallel with other child processes.

7. It serializes the result back into binary data.

8. It copies that binary data back through the socket.

9. It deserializes the binary data back into Python objects in the parent process.

10. It merges the results from multiple children into a single `list` to return.

Although it looks simple to the programmer, the `multiprocessing` module and `ProcessPoolExecutor` class do a huge amount of work to make parallelism possible. In most other languages, the only touch point you need to coordinate two threads is a single lock or atomic operation (see Item 54: "Use Lock to Prevent Data Races in Threads" for an example). The overhead of using `multiprocessing` via `ProcessPoolExecutor` is high because of all of the serialization and deserialization that must happen between the parent and child processes.

This scheme is well suited to certain types of isolated, high-leverage tasks. By *isolated*, I mean functions that don't need to share state with other parts of the program. By *high-leverage tasks*, I mean situations in which only a small amount of data must be transferred between the parent and child processes to enable a large amount of computation. The greatest common divisor algorithm is one example of this, but many other mathematical algorithms work similarly.

If your computation doesn't have these characteristics, then the overhead of `ProcessPoolExecutor` may prevent it from speeding up your program through parallelization. When that happens, `multiprocessing` provides more advanced facilities for shared memory, cross-process locks, queues, and proxies. But all of these features are very complex. It's hard enough to reason about such tools in the memory space of a single process shared between Python threads. Extending that complexity to other processes and involving sockets makes this much more difficult to understand.

I suggest that you initially avoid all parts of the `multiprocessing` built-in module. You can start by using the `ThreadPoolExecutor` class to run isolated, high-leverage functions in threads. Later you can move to the `ProcessPoolExecutor` to get a speedup. Finally, when you've completely exhausted the other options, you can consider using the `multiprocessing` module directly.

Things to Remember

✦ Moving CPU bottlenecks to C-extension modules can be an effective way to improve performance while maximizing your investment in Python code. However, doing so has a high cost and may introduce bugs.

✦ The `multiprocessing` module provides powerful tools that can parallelize certain types of Python computation with minimal effort.

✦ The power of `multiprocessing` is best accessed through the `concurrent.futures` built-in module and its simple `ProcessPoolExecutor` class.

✦ Avoid the advanced (and complicated) parts of the `multiprocessing` module until you've exhausted all other options.

Chapter 8

Robustness and Performance

Once you've written a useful Python program, the next step is to *productionize* your code so it's bulletproof. Making programs dependable when they encounter unexpected circumstances is just as important as making programs with correct functionality. Python has built-in features and modules that aid in hardening your programs so they are robust in a wide variety of situations.

One dimension of robustness is scalability and performance. When you're implementing Python programs that handle a non-trivial amount of data, you'll often see slowdowns caused by the algorithmic complexity of your code or other types of computational overhead. Luckily, Python includes many of the algorithms and data structures you need to achieve high performance with minimal effort.

Item 65: Take Advantage of Each Block in try/except /else/finally

There are four distinct times when you might want to take action during exception handling in Python. These are captured in the functionality of try, except, else, and finally blocks. Each block serves a unique purpose in the compound statement, and their various combinations are useful (see Item 87: "Define a Root Exception to Insulate Callers from APIs" for another example).

finally Blocks

Use try/finally when you want exceptions to propagate up but also want to run cleanup code even when exceptions occur. One common usage of try/finally is for reliably closing file handles (see Item 66: "Consider contextlib and with Statements for Reusable try/finally Behavior" for another—likely better—approach):

```
def try_finally_example(filename):
    print('* Opening file')
```

```
    handle = open(filename, encoding='utf-8') # Maybe OSError
    try:
        print('* Reading data')
        return handle.read()  # Maybe UnicodeDecodeError
    finally:
        print('* Calling close()')
        handle.close()              # Always runs after try block
```

Any exception raised by the read method will always propagate up to the calling code, but the close method of handle in the finally block will run first:

```
filename = 'random_data.txt'

with open(filename, 'wb') as f:
    f.write(b'\xf1\xf2\xf3\xf4\xf5')   # Invalid utf-8

data = try_finally_example(filename)

>>>
* Opening file
* Reading data
* Calling close()
Traceback ...
UnicodeDecodeError: 'utf-8' codec can't decode byte 0xf1 in
➥position 0: invalid continuation byte
```

You must call open before the try block because exceptions that occur when opening the file (like OSError if the file does not exist) should skip the finally block entirely:

```
try_finally_example('does_not_exist.txt')

>>>
* Opening file
Traceback ...
FileNotFoundError: [Errno 2] No such file or directory:
➥'does_not_exist.txt'
```

else Blocks

Use try/except/else to make it clear which exceptions will be handled by your code and which exceptions will propagate up. When the try block doesn't raise an exception, the else block runs. The else block helps you minimize the amount of code in the try block, which is good for isolating potential exception causes and improves readability. For example, say that I want to load JSON dictionary data from a string and return the value of a key it contains:

```
import json

def load_json_key(data, key):
    try:
        print('* Loading JSON data')
        result_dict = json.loads(data)  # May raise ValueError
    except ValueError as e:
        print('* Handling ValueError')
        raise KeyError(key) from e
    else:
        print('* Looking up key')
        return result_dict[key]          # May raise KeyError
```

In the successful case, the JSON data is decoded in the try block, and then the key lookup occurs in the else block:

```
assert load_json_key('{"foo": "bar"}', 'foo') == 'bar'

>>>
* Loading JSON data
* Looking up key
```

If the input data isn't valid JSON, then decoding with json.loads raises a ValueError. The exception is caught by the except block and handled:

```
load_json_key('{"foo": bad payload', 'foo')

>>>
* Loading JSON data
* Handling ValueError
Traceback ...
JSONDecodeError: Expecting value: line 1 column 9 (char 8)

The above exception was the direct cause of the following
➥exception:

Traceback ...
KeyError: 'foo'
```

If the key lookup raises any exceptions, they propagate up to the caller because they are outside the try block. The else clause ensures that what follows the try/except is visually distinguished from the except block. This makes the exception propagation behavior clear:

```
load_json_key('{"foo": "bar"}', 'does not exist')
```

```
>>>
* Loading JSON data
* Looking up key
Traceback ...
KeyError: 'does not exist'
```

Everything Together

Use try/except/else/finally when you want to do it all in one compound statement. For example, say that I want to read a description of work to do from a file, process it, and then update the file in-place. Here, the try block is used to read the file and process it; the except block is used to handle exceptions from the try block that are expected; the else block is used to update the file in place and allow related exceptions to propagate up; and the finally block cleans up the file handle:

```
UNDEFINED = object()

def divide_json(path):
    print('* Opening file')
    handle = open(path, 'r+')    # May raise OSError
    try:
        print('* Reading data')
        data = handle.read()     # May raise UnicodeDecodeError
        print('* Loading JSON data')
        op = json.loads(data)    # May raise ValueError
        print('* Performing calculation')
        value = (
            op['numerator'] /
            op['denominator'])   # May raise ZeroDivisionError
    except ZeroDivisionError as e:
        print('* Handling ZeroDivisionError')
        return UNDEFINED
    else:
        print('* Writing calculation')
        op['result'] = value
        result = json.dumps(op)
        handle.seek(0)           # May raise OSError
        handle.write(result)     # May raise OSError
        return value
    finally:
        print('* Calling close()')
        handle.close()           # Always runs
```

In the successful case, the try, else, and finally blocks run:

```
temp_path = 'random_data.json'

with open(temp_path, 'w') as f:
    f.write('{"numerator": 1, "denominator": 10}')

assert divide_json(temp_path) == 0.1

>>>
* Opening file
* Reading data
* Loading JSON data
* Performing calculation
* Writing calculation
* Calling close()
```

If the calculation is invalid, the try, except, and finally blocks run, but the else block does not:

```
with open(temp_path, 'w') as f:
    f.write('{"numerator": 1, "denominator": 0}')

assert divide_json(temp_path) is UNDEFINED

>>>
* Opening file
* Reading data
* Loading JSON data
* Performing calculation
* Handling ZeroDivisionError
* Calling close()
```

If the JSON data was invalid, the try block runs and raises an exception, the finally block runs, and then the exception is propagated up to the caller. The except and else blocks do not run:

```
with open(temp_path, 'w') as f:
    f.write('{"numerator": 1 bad data')

divide_json(temp_path)

>>>
* Opening file
* Reading data
* Loading JSON data
* Calling close()
Traceback ...
JSONDecodeError: Expecting ',' delimiter: line 1 column 17
➥(char 16)
```

This layout is especially useful because all of the blocks work together in intuitive ways. For example, here I simulate this by running the divide_json function at the same time that my hard drive runs out of disk space:

```
with open(temp_path, 'w') as f:
    f.write('{"numerator": 1, "denominator": 10}')

divide_json(temp_path)

>>>
* Opening file
* Reading data
* Loading JSON data
* Performing calculation
* Writing calculation
* Calling close()
Traceback ...
OSError: [Errno 28] No space left on device
```

When the exception was raised in the else block while rewriting the result data, the finally block still ran and closed the file handle as expected.

Things to Remember

✦ The try/finally compound statement lets you run cleanup code regardless of whether exceptions were raised in the try block.

✦ The else block helps you minimize the amount of code in try blocks and visually distinguish the success case from the try/except blocks.

✦ An else block can be used to perform additional actions after a successful try block but before common cleanup in a finally block.

Item 66: Consider contextlib and with Statements for Reusable try/finally Behavior

The with statement in Python is used to indicate when code is running in a special context. For example, mutual-exclusion locks (see Item 54: "Use Lock to Prevent Data Races in Threads") can be used in with statements to indicate that the indented code block runs only while the lock is held:

```
from threading import Lock

lock = Lock()
```

```
with lock:
    # Do something while maintaining an invariant
    ...
```

The example above is equivalent to this try/finally construction because the Lock class properly enables the with statement (see Item 65: "Take Advantage of Each Block in try/except/else/finally" for more about try/finally):

```
lock.acquire()
try:
    # Do something while maintaining an invariant
    ...
finally:
    lock.release()
```

The with statement version of this is better because it eliminates the need to write the repetitive code of the try/finally construction, and it ensures that you don't forget to have a corresponding release call for every acquire call.

It's easy to make your objects and functions work in with statements by using the contextlib built-in module. This module contains the contextmanager decorator (see Item 26: "Define Function Decorators with functools.wraps" for background), which lets a simple function be used in with statements. This is much easier than defining a new class with the special methods __enter__ and __exit__ (the standard way).

For example, say that I want a region of code to have more debug logging sometimes. Here, I define a function that does logging at two severity levels:

```
import logging

def my_function():
    logging.debug('Some debug data')
    logging.error('Error log here')
    logging.debug('More debug data')
```

The default log level for my program is WARNING, so only the error message will print to screen when I run the function:

```
my_function()

>>>
Error log here
```

I can elevate the log level of this function temporarily by defining a context manager. This helper function boosts the logging severity level before running the code in the with block and reduces the logging severity level afterward:

```
from contextlib import contextmanager
```

```
@contextmanager
def debug_logging(level):
    logger = logging.getLogger()
    old_level = logger.getEffectiveLevel()
    logger.setLevel(level)
    try:
        yield
    finally:
        logger.setLevel(old_level)
```

The yield expression is the point at which the with block's contents will execute (see Item 30: "Consider Generators Instead of Returning Lists" for background). Any exceptions that happen in the with block will be re-raised by the yield expression for you to catch in the helper function (see Item 35: "Avoid Causing State Transitions in Generators with throw" for how that works).

Now, I can call the same logging function again but in the debug_logging context. This time, all of the debug messages are printed to the screen during the with block. The same function running outside the with block won't print debug messages:

```
with debug_logging(logging.DEBUG):
    print('* Inside:')
    my_function()
```

```
print('* After:')
my_function()
```

```
>>>
* Inside:
Some debug data
Error log here
More debug data
* After:
Error log here
```

Using with Targets

The context manager passed to a with statement may also return an object. This object is assigned to a local variable in the as part of the compound statement. This gives the code running in the with block the ability to directly interact with its context.

For example, say I want to write a file and ensure that it's always closed correctly. I can do this by passing open to the with statement. open returns a file handle for the as target of with, and it closes the handle when the with block exits:

```
with open('my_output.txt', 'w') as handle:
    handle.write('This is some data!')
```

This approach is more Pythonic than manually opening and closing the file handle every time. It gives you confidence that the file is eventually closed when execution leaves the with statement. By highlighting the critical section, it also encourages you to reduce the amount of code that executes while the file handle is open, which is good practice in general.

To enable your own functions to supply values for as targets, all you need to do is yield a value from your context manager. For example, here I define a context manager to fetch a Logger instance, set its level, and then yield it as the target:

```
@contextmanager
def log_level(level, name):
    logger = logging.getLogger(name)
    old_level = logger.getEffectiveLevel()
    logger.setLevel(level)
    try:
        yield logger
    finally:
        logger.setLevel(old_level)
```

Calling logging methods like debug on the as target produces output because the logging severity level is set low enough in the with block on that specific Logger instance. Using the logging module directly won't print anything because the default logging severity level for the default program logger is WARNING:

```
with log_level(logging.DEBUG, 'my-log') as logger:
    logger.debug(f'This is a message for {logger.name}!')
    logging.debug('This will not print')

>>>
This is a message for my-log!
```

After the with statement exits, calling debug logging methods on the Logger named 'my-log' will not print anything because the default

logging severity level has been restored. Error log messages will
always print:

```
logger = logging.getLogger('my-log')
logger.debug('Debug will not print')
logger.error('Error will print')
```

```
>>>
Error will print
```

Later, I can change the name of the logger I want to use by simply
updating the with statement. This will point the Logger that's the as
target in the with block to a different instance, but I won't have to
update any of my other code to match:

```
with log_level(logging.DEBUG, 'other-log') as logger:
    logger.debug(f'This is a message for {logger.name}!')
    logging.debug('This will not print')
```

```
>>>
This is a message for other-log!
```

This isolation of state and decoupling between creating a context and
acting within that context is another benefit of the with statement.

Things to Remember

✦ The with statement allows you to reuse logic from try/finally blocks
and reduce visual noise.

✦ The contextlib built-in module provides a contextmanager decorator
that makes it easy to use your own functions in with statements.

✦ The value yielded by context managers is supplied to the as part
of the with statement. It's useful for letting your code directly access
the cause of a special context.

Item 67: Use datetime Instead of time for Local Clocks

Coordinated Universal Time (UTC) is the standard, time-zone-
independent representation of time. UTC works great for computers
that represent time as seconds since the UNIX epoch. But UTC isn't
ideal for humans. Humans reference time relative to where they're
currently located. People say "noon" or "8 am" instead of "UTC 15:00
minus 7 hours." If your program handles time, you'll probably find
yourself converting time between UTC and local clocks for the sake of
human understanding.

Python provides two ways of accomplishing time zone conversions. The old way, using the time built-in module, is terribly error prone. The new way, using the datetime built-in module, works great with some help from the community-built package named pytz.

You should be acquainted with both time and datetime to thoroughly understand why datetime is the best choice and time should be avoided.

The time Module

The localtime function from the time built-in module lets you convert a UNIX timestamp (seconds since the UNIX epoch in UTC) to a local time that matches the host computer's time zone (Pacific Daylight Time in my case). This local time can be printed in human-readable format using the strftime function:

```
import time

now = 1552774475
local_tuple = time.localtime(now)
time_format = '%Y-%m-%d %H:%M:%S'
time_str = time.strftime(time_format, local_tuple)
print(time_str)

>>>
2019-03-16 15:14:35
```

You'll often need to go the other way as well, starting with user input in human-readable local time and converting it to UTC time. You can do this by using the strptime function to parse the time string, and then calling mktime to convert local time to a UNIX timestamp:

```
time_tuple = time.strptime(time_str, time_format)
utc_now = time.mktime(time_tuple)
print(utc_now)

>>>
1552774475.0
```

How do you convert local time in one time zone to local time in another time zone? For example, say that I'm taking a flight between San Francisco and New York, and I want to know what time it will be in San Francisco when I've arrived in New York.

I might initially assume that I can directly manipulate the return values from the time, localtime, and strptime functions to do time zone conversions. But this is a very bad idea. Time zones change all the time due to local laws. It's too complicated to manage yourself, especially if you want to handle every global city for flight departures and arrivals.

Many operating systems have configuration files that keep up with the time zone changes automatically. Python lets you use these time zones through the time module if your platform supports it. On other platforms, such as Windows, some time zone functionality isn't available from time at all. For example, here I parse a departure time from the San Francisco time zone, Pacific Daylight Time (PDT):

```
import os

if os.name == 'nt':
    print("This example doesn't work on Windows")
else:
    parse_format = '%Y-%m-%d %H:%M:%S %Z'
    depart_sfo = '2019-03-16 15:45:16 PDT'
    time_tuple = time.strptime(depart_sfo, parse_format)
    time_str = time.strftime(time_format, time_tuple)
    print(time_str)
```

```
>>>
2019-03-16 15:45:16
```

After seeing that 'PDT' works with the strptime function, I might also assume that other time zones known to my computer will work. Unfortunately, this isn't the case. strptime raises an exception when it sees Eastern Daylight Time (EDT), which is the time zone for New York:

```
arrival_nyc = '2019-03-16 23:33:24 EDT'
time_tuple = time.strptime(arrival_nyc, time_format)
```

```
>>>
Traceback ...
ValueError: unconverted data remains:   EDT
```

The problem here is the platform-dependent nature of the time module. Its behavior is determined by how the underlying C functions work with the host operating system. This makes the functionality of the time module unreliable in Python. The time module fails to consistently work properly for multiple local times. Thus, you should avoid using the time module for this purpose. If you must use time, use it only to convert between UTC and the host computer's local time. For all other types of conversions, use the datetime module.

The datetime Module

The second option for representing times in Python is the datetime class from the datetime built-in module. Like the time module, datetime can be used to convert from the current time in UTC to local time.

Here, I convert the present time in UTC to my computer's local time, PDT:

```
from datetime import datetime, timezone

now = datetime(2019, 3, 16, 22, 14, 35)
now_utc = now.replace(tzinfo=timezone.utc)
now_local = now_utc.astimezone()
print(now_local)
```

```
>>>
2019-03-16 15:14:35-07:00
```

The datetime module can also easily convert a local time back to a UNIX timestamp in UTC:

```
time_str = '2019-03-16 15:14:35'
now = datetime.strptime(time_str, time_format)
time_tuple = now.timetuple()
utc_now = time.mktime(time_tuple)
print(utc_now)
```

```
>>>
1552774475.0
```

Unlike the time module, the datetime module has facilities for reliably converting from one local time to another local time. However, datetime only provides the machinery for time zone operations with its tzinfo class and related methods. The Python default installation is missing time zone definitions besides UTC.

Luckily, the Python community has addressed this gap with the pytz module that's available for download from the Python Package Index (see Item 82: "Know Where to Find Community-Built Modules" for how to install it). pytz contains a full database of every time zone definition you might need.

To use pytz effectively, you should always convert local times to UTC first. Perform any datetime operations you need on the UTC values (such as offsetting). Then, convert to local times as a final step.

For example, here I convert a New York City flight arrival time to a UTC datetime. Although some of these calls seem redundant, all of them are necessary when using pytz:

```
import pytz

arrival_nyc = '2019-03-16 23:33:24'
nyc_dt_naive = datetime.strptime(arrival_nyc, time_format)
```

```
eastern = pytz.timezone('US/Eastern')
nyc_dt = eastern.localize(nyc_dt_naive)
utc_dt = pytz.utc.normalize(nyc_dt.astimezone(pytz.utc))
print(utc_dt)
```

```
>>>
2019-03-17 03:33:24+00:00
```

Once I have a UTC datetime, I can convert it to San Francisco local time:

```
pacific = pytz.timezone('US/Pacific')
sf_dt = pacific.normalize(utc_dt.astimezone(pacific))
print(sf_dt)
```

```
>>>
2019-03-16 20:33:24-07:00
```

Just as easily, I can convert it to the local time in Nepal:

```
nepal = pytz.timezone('Asia/Katmandu')
nepal_dt = nepal.normalize(utc_dt.astimezone(nepal))
print(nepal_dt)
```

```
>>>
2019-03-17 09:18:24+05:45
```

With datetime and pytz, these conversions are consistent across all environments, regardless of what operating system the host computer is running.

Things to Remember

✦ Avoid using the time module for translating between different time zones.

✦ Use the datetime built-in module along with the pytz community module to reliably convert between times in different time zones.

✦ Always represent time in UTC and do conversions to local time as the very final step before presentation.

Item 68: Make pickle Reliable with copyreg

The pickle built-in module can serialize Python objects into a stream of bytes and deserialize bytes back into objects. Pickled byte streams shouldn't be used to communicate between untrusted parties. The purpose of pickle is to let you pass Python objects between programs that you control over binary channels.

Note

The pickle module's serialization format is unsafe by design. The serialized data contains what is essentially a program that describes how to reconstruct the original Python object. This means a malicious pickle payload could be used to compromise any part of a Python program that attempts to deserialize it.

In contrast, the json module is safe by design. Serialized JSON data contains a simple description of an object hierarchy. Deserializing JSON data does not expose a Python program to additional risk. Formats like JSON should be used for communication between programs or people who don't trust each other.

For example, say that I want to use a Python object to represent the state of a player's progress in a game. The game state includes the level the player is on and the number of lives they have remaining:

```python
class GameState:
    def __init__(self):
        self.level = 0
        self.lives = 4
```

The program modifies this object as the game runs:

```python
state = GameState()
state.level += 1  # Player beat a level
state.lives -= 1  # Player had to try again

print(state.__dict__)
```

```
>>>
{'level': 1, 'lives': 3}
```

When the user quits playing, the program can save the state of the game to a file so it can be resumed at a later time. The pickle module makes it easy to do this. Here, I use the dump function to write the GameState object to a file:

```python
import pickle

state_path = 'game_state.bin'
with open(state_path, 'wb') as f:
    pickle.dump(state, f)
```

Later, I can call the load function with the file and get back the GameState object as if it had never been serialized:

```python
with open(state_path, 'rb') as f:
    state_after = pickle.load(f)

print(state_after.__dict__)
```

```
>>>
{'level': 1, 'lives': 3}
```

The problem with this approach is what happens as the game's features expand over time. Imagine that I want the player to earn points toward a high score. To track the player's points, I'd add a new field to the GameState class

```
class GameState:
    def __init__(self):
        self.level = 0
        self.lives = 4
        self.points = 0   # New field
```

Serializing the new version of the GameState class using pickle will work exactly as before. Here, I simulate the round-trip through a file by serializing to a string with dumps and back to an object with loads:

```
state = GameState()
serialized = pickle.dumps(state)
state_after = pickle.loads(serialized)
print(state_after.__dict__)
```

```
>>>
{'level': 0, 'lives': 4, 'points': 0}
```

But what happens to older saved GameState objects that the user may want to resume? Here, I unpickle an old game file by using a program with the new definition of the GameState class:

```
with open(state_path, 'rb') as f:
    state_after = pickle.load(f)

print(state_after.__dict__)
```

```
>>>
{'level': 1, 'lives': 3}
```

The points attribute is missing! This is especially confusing because the returned object is an instance of the new GameState class:

```
assert isinstance(state_after, GameState)
```

This behavior is a byproduct of the way the pickle module works. Its primary use case is making object serialization easy. As soon as your use of pickle moves beyond trivial usage, the module's functionality starts to break down in surprising ways.

Fixing these problems is straightforward using the copyreg built-in module. The copyreg module lets you register the functions responsible for serializing and deserializing Python objects, allowing you to control the behavior of pickle and make it more reliable.

Default Attribute Values

In the simplest case, you can use a constructor with default arguments (see Item 23: "Provide Optional Behavior with Keyword Arguments" for background) to ensure that GameState objects will always have all attributes after unpickling. Here, I redefine the constructor this way:

```
class GameState:
    def __init__(self, level=0, lives=4, points=0):
        self.level = level
        self.lives = lives
        self.points = points
```

To use this constructor for pickling, I define a helper function that takes a GameState object and turns it into a tuple of parameters for the copyreg module. The returned tuple contains the function to use for unpickling and the parameters to pass to the unpickling function:

```
def pickle_game_state(game_state):
    kwargs = game_state.__dict__
    return unpickle_game_state, (kwargs,)
```

Now, I need to define the unpickle_game_state helper. This function takes serialized data and parameters from pickle_game_state and returns the corresponding GameState object. It's a tiny wrapper around the constructor:

```
def unpickle_game_state(kwargs):
    return GameState(**kwargs)
```

Now, I register these functions with the copyreg built-in module:

```
import copyreg

copyreg.pickle(GameState, pickle_game_state)
```

After registration, serializing and deserializing works as before:

```
state = GameState()
state.points += 1000
serialized = pickle.dumps(state)
state_after = pickle.loads(serialized)
print(state_after.__dict__)

>>>
{'level': 0, 'lives': 4, 'points': 1000}
```

With this registration done, now I'll change the definition of GameState again to give the player a count of magic spells to use. This change is similar to when I added the points field to GameState:

```
class GameState:
    def __init__(self, level=0, lives=4, points=0, magic=5):
        self.level = level
        self.lives = lives
        self.points = points
        self.magic = magic  # New field
```

But unlike before, deserializing an old GameState object will result in valid game data instead of missing attributes. This works because unpickle_game_state calls the GameState constructor directly instead of using the pickle module's default behavior of saving and restoring only the attributes that belong to an object. The GameState constructor's keyword arguments have default values that will be used for any parameters that are missing. This causes old game state files to receive the default value for the new magic field when they are deserialized:

```
print('Before:', state.__dict__)
state_after = pickle.loads(serialized)
print('After: ', state_after.__dict__)

>>>
Before: {'level': 0, 'lives': 4, 'points': 1000}
After:  {'level': 0, 'lives': 4, 'points': 1000, 'magic': 5}
```

Versioning Classes

Sometimes you need to make backward-incompatible changes to your Python objects by removing fields. Doing so prevents the default argument approach above from working.

For example, say I realize that a limited number of lives is a bad idea, and I want to remove the concept of lives from the game. Here, I redefine the GameState class to no longer have a lives field:

```
class GameState:
    def __init__(self, level=0, points=0, magic=5):
        self.level = level
        self.points = points
        self.magic = magic
```

The problem is that this breaks deserialization of old game data. All fields from the old data, even ones removed from the class, will be passed to the GameState constructor by the unpickle_game_state function:

```
pickle.loads(serialized)
```

```
>>>
Traceback ...
TypeError: __init__() got an unexpected keyword argument
➥'lives'
```

I can fix this by adding a version parameter to the functions supplied to copyreg. New serialized data will have a version of 2 specified when pickling a new GameState object:

```
def pickle_game_state(game_state):
    kwargs = game_state.__dict__
    kwargs['version'] = 2
    return unpickle_game_state, (kwargs,)
```

Old versions of the data will not have a version argument present, which means I can manipulate the arguments passed to the GameState constructor accordingly:

```
def unpickle_game_state(kwargs):
    version = kwargs.pop('version', 1)
    if version == 1:
        del kwargs['lives']
    return GameState(**kwargs)
```

Now, deserializing an old object works properly:

```
copyreg.pickle(GameState, pickle_game_state)
print('Before:', state.__dict__)
state_after = pickle.loads(serialized)
print('After: ', state_after.__dict__)
```

```
>>>
Before: {'level': 0, 'lives': 4, 'points': 1000}
After:  {'level': 0, 'points': 1000, 'magic': 5}
```

I can continue using this approach to handle changes between future versions of the same class. Any logic I need to adapt an old version of the class to a new version of the class can go in the unpickle_game_state function.

Stable Import Paths

One other issue you may encounter with pickle is breakage from renaming a class. Often over the life cycle of a program, you'll refactor your code by renaming classes and moving them to other modules. Unfortunately, doing so breaks the pickle module unless you're careful.

Here, I rename the GameState class to BetterGameState and remove the old class from the program entirely:

```
class BetterGameState:
    def __init__(self, level=0, points=0, magic=5):
        self.level = level
        self.points = points
        self.magic = magic
```

Attempting to deserialize an old GameState object now fails because the class can't be found:

```
pickle.loads(serialized)
```

```
>>>
Traceback ...
AttributeError: Can't get attribute 'GameState' on <module
➡'__main__' from 'my_code.py'>
```

The cause of this exception is that the import path of the serialized object's class is encoded in the pickled data:

```
print(serialized)
```

```
>>>
b'\x80\x04\x95A\x00\x00\x00\x00\x00\x00\x00\x8c\x08__main__
➡\x94\x8c\tGameState\x94\x93\x94)\x81\x94}\x94(\x8c\x05level
➡\x94K\x00\x8c\x06points\x94K\x00\x8c\x05magic\x94K\x05ub.'
```

The solution is to use copyreg again. I can specify a stable identifier for the function to use for unpickling an object. This allows me to transition pickled data to different classes with different names when it's deserialized. It gives me a level of indirection:

```
copyreg.pickle(BetterGameState, pickle_game_state)
```

After I use copyreg, you can see that the import path to unpickle_game_state is encoded in the serialized data instead of BetterGameState:

```
state = BetterGameState()
serialized = pickle.dumps(state)
print(serialized)
```

```
>>>
b'\x80\x04\x95W\x00\x00\x00\x00\x00\x00\x00\x8c\x08__main__
➡\x94\x8c\x13unpickle_game_state\x94\x93\x94}\x94(\x8c
➡\x05level\x94K\x00\x8c\x06points\x94K\x00\x8c\x05magic\x94K
➡\x05\x8c\x07version\x94K\x02u\x85\x94R\x94.'
```

The only gotcha is that I can't change the path of the module in which the `unpickle_game_state` function is present. Once I serialize data with a function, it must remain available on that import path for deserialization in the future.

Things to Remember

✦ The `pickle` built-in module is useful only for serializing and deserializing objects between trusted programs.

✦ Deserializing previously pickled objects may break if the classes involved have changed over time (e.g., attributes have been added or removed).

✦ Use the `copyreg` built-in module with `pickle` to ensure backward compatibility for serialized objects.

Item 69: Use decimal **When Precision Is Paramount**

Python is an excellent language for writing code that interacts with numerical data. Python's integer type can represent values of any practical size. Its double-precision floating point type complies with the IEEE 754 standard. The language also provides a standard complex number type for imaginary values. However, these aren't enough for every situation.

For example, say that I want to compute the amount to charge a customer for an international phone call. I know the time in minutes and seconds that the customer was on the phone (say, 3 minutes 42 seconds). I also have a set rate for the cost of calling Antarctica from the United States ($1.45/minute). What should the charge be?

With floating point math, the computed charge seems reasonable

```
rate = 1.45
seconds = 3*60 + 42
cost = rate * seconds / 60
print(cost)
```

```
>>>
5.364999999999999
```

The result is 0.0001 short of the correct value (5.365) due to how IEEE 754 floating point numbers are represented. I might want to round up this value to 5.37 to properly cover all costs incurred by the customer. However, due to floating point error, rounding to the nearest whole cent actually reduces the final charge (from 5.364 to 5.36) instead of increasing it (from 5.365 to 5.37):

```
print(round(cost, 2))
```

```
>>>
5.36
```

The solution is to use the Decimal class from the decimal built-in module. The Decimal class provides fixed point math of 28 decimal places by default. It can go even higher, if required. This works around the precision issues in IEEE 754 floating point numbers. The class also gives you more control over rounding behaviors.

For example, redoing the Antarctica calculation with Decimal results in the exact expected charge instead of an approximation:

```
from decimal import Decimal

rate = Decimal('1.45')
seconds = Decimal(3*60 + 42)
cost = rate * seconds / Decimal(60)
print(cost)
```

```
>>>
5.365
```

Decimal instances can be given starting values in two different ways. The first way is by passing a str containing the number to the Decimal constructor. This ensures that there is no loss of precision due to the inherent nature of Python floating point numbers. The second way is by directly passing a float or an int instance to the constructor. Here, you can see that the two construction methods result in different behavior.

```
print(Decimal('1.45'))
print(Decimal(1.45))
```

```
>>>
1.45
1.4499999999999999555910790149937383838305473327637863671875
```

The same problem doesn't happen if I supply integers to the Decimal constructor:

```
print('456')
print(456)
```

```
>>>
456
456
```

If you care about exact answers, err on the side of caution and use the str constructor for the Decimal type.

Getting back to the phone call example, say that I also want to support very short phone calls between places that are much cheaper to connect (like Toledo and Detroit). Here, I compute the charge for a phone call that was 5 seconds long with a rate of $0.05/minute:

```
rate = Decimal('0.05')
seconds = Decimal('5')
small_cost = rate * seconds / Decimal(60)
print(small_cost)
```

```
>>>
0.004166666666666666666666666667
```

The result is so low that it is decreased to zero when I try to round it to the nearest whole cent. This won't do!

```
print(round(small_cost, 2))
```

```
>>>
0.00
```

Luckily, the Decimal class has a built-in function for rounding to exactly the decimal place needed with the desired rounding behavior. This works for the higher cost case from earlier:

```
from decimal import ROUND_UP

rounded = cost.quantize(Decimal('0.01'), rounding=ROUND_UP)
print(f'Rounded {cost} to {rounded}')
```

```
>>>
Rounded 5.365 to 5.37
```

Using the quantize method this way also properly handles the small usage case for short, cheap phone calls:.

```
rounded = small_cost.quantize(Decimal('0.01'),
                              rounding=ROUND_UP)
print(f'Rounded {small_cost} to {rounded}')
```

```
>>>
Rounded 0.004166666666666666666666666667 to 0.01
```

While Decimal works great for fixed point numbers, it still has limitations in its precision (e.g., 1/3 will be an approximation). For representing rational numbers with no limit to precision, consider using the Fraction class from the fractions built-in module.

Things to Remember

✦ Python has built-in types and classes in modules that can represent practically every type of numerical value.

✦ The Decimal class is ideal for situations that require high precision and control over rounding behavior, such as computations of monetary values.

✦ Pass str instances to the Decimal constructor instead of float instances if it's important to compute exact answers and not floating point approximations.

Item 70: Profile Before Optimizing

The dynamic nature of Python causes surprising behaviors in its runtime performance. Operations you might assume would be slow are actually very fast (e.g., string manipulation, generators). Language features you might assume would be fast are actually very slow (e.g., attribute accesses, function calls). The true source of slowdowns in a Python program can be obscure.

The best approach is to ignore your intuition and directly measure the performance of a program before you try to optimize it. Python provides a built-in *profiler* for determining which parts of a program are responsible for its execution time. This means you can focus your optimization efforts on the biggest sources of trouble and ignore parts of the program that don't impact speed (i.e., follow Amdahl's law).

For example, say that I want to determine why an algorithm in a program is slow. Here, I define a function that sorts a list of data using an insertion sort:

```
def insertion_sort(data):
    result = []
    for value in data:
        insert_value(result, value)
    return result
```

The core mechanism of the insertion sort is the function that finds the insertion point for each piece of data. Here, I define an extremely inefficient version of the insert_value function that does a linear scan over the input array:

```
def insert_value(array, value):
    for i, existing in enumerate(array):
        if existing > value:
            array.insert(i, value)
```

```
        return
    array.append(value)
```

To profile `insertion_sort` and `insert_value`, I create a data set of random numbers and define a `test` function to pass to the profiler:

```
from random import randint

max_size = 10**4
data = [randint(0, max_size) for _ in range(max_size)]
test = lambda: insertion_sort(data)
```

Python provides two built-in profilers: one that is pure Python (profile) and another that is a C-extension module (cProfile). The cProfile built-in module is better because of its minimal impact on the performance of your program while it's being profiled. The pure-Python alternative imposes a high overhead that skews the results.

> Note
> When profiling a Python program, be sure that what you're measuring is the code itself and not external systems. Beware of functions that access the network or resources on disk. These may appear to have a large impact on your program's execution time because of the slowness of the underlying systems. If your program uses a cache to mask the latency of slow resources like these, you should ensure that it's properly warmed up before you start profiling.

Here, I instantiate a `Profile` object from the `cProfile` module and run the test function through it using the `runcall` method:

```
from cProfile import Profile

profiler = Profile()
profiler.runcall(test)
```

When the test function has finished running, I can extract statistics about its performance by using the `pstats` built-in module and its `Stats` class. Various methods on a `Stats` object adjust how to select and sort the profiling information to show only the things I care about:

```
from pstats import Stats

stats = Stats(profiler)
stats.strip_dirs()
stats.sort_stats('cumulative')
stats.print_stats()
```

The output is a table of information organized by function. The data sample is taken only from the time the profiler was active, during the `runcall` method above:

```
>>>
        20003 function calls in 1.320 seconds

  Ordered by: cumulative time

  ncalls  tottime  percall  cumtime  percall filename:lineno(function)
       1    0.000    0.000    1.320    1.320 main.py:35(<lambda>)
       1    0.003    0.003    1.320    1.320 main.py:10(insertion_sort)
   10000    1.306    0.000    1.317    0.000 main.py:20(insert_value)
    9992    0.011    0.000    0.011    0.000 {method 'insert' of 'list' objects}
       8    0.000    0.000    0.000    0.000 {method 'append' of 'list' objects}
```

Here's a quick guide to what the profiler statistics columns mean:

- `ncalls`: The number of calls to the function during the profiling period.

- `tottime`: The number of seconds spent executing the function, excluding time spent executing other functions it calls.

- `tottime percall`: The average number of seconds spent in the function each time it is called, excluding time spent executing other functions it calls. This is `tottime` divided by `ncalls`.

- `cumtime`: The cumulative number of seconds spent executing the function, including time spent in all other functions it calls.

- `cumtime percall`: The average number of seconds spent in the function each time it is called, including time spent in all other functions it calls. This is `cumtime` divided by `ncalls`.

Looking at the profiler statistics table above, I can see that the biggest use of CPU in my test is the cumulative time spent in the `insert_value` function. Here, I redefine that function to use the `bisect` built-in module (see Item 72: "Consider Searching Sorted Sequences with `bisect`"):

```
from bisect import bisect_left

def insert_value(array, value):
    i = bisect_left(array, value)
    array.insert(i, value)
```

I can run the profiler again and generate a new table of profiler statistics. The new function is much faster, with a cumulative time spent that is nearly 100 times smaller than with the previous `insert_value` function:

```
>>>
        30003 function calls in 0.017 seconds

   Ordered by: cumulative time

   ncalls  tottime  percall  cumtime  percall filename:lineno(function)
        1    0.000    0.000    0.017    0.017 main.py:35(<lambda>)
        1    0.002    0.002    0.017    0.017 main.py:10(insertion_sort)
    10000    0.003    0.000    0.015    0.000 main.py:110(insert_value)
    10000    0.008    0.000    0.008    0.000 {method 'insert' of 'list' objects}
    10000    0.004    0.000    0.004    0.000 {built-in method _bisect.bisect_left}
```

Sometimes when you're profiling an entire program, you might find that a common utility function is responsible for the majority of execution time. The default output from the profiler makes such a situation difficult to understand because it doesn't show that the utility function is called by many different parts of your program.

For example, here the `my_utility` function is called repeatedly by two different functions in the program:

```python
def my_utility(a, b):
    c = 1
    for i in range(100):
        c += a * b

def first_func():
    for _ in range(1000):
        my_utility(4, 5)

def second_func():
    for _ in range(10):
        my_utility(1, 3)

def my_program():
    for _ in range(20):
        first_func()
        second_func()
```

Profiling this code and using the default `print_stats` output generates statistics that are confusing:

```
>>>
        20242 function calls in 0.118 seconds

   Ordered by: cumulative time

   ncalls  tottime  percall  cumtime  percall filename:lineno(function)
        1    0.000    0.000    0.118    0.118 main.py:176(my_program)
       20    0.003    0.000    0.117    0.006 main.py:168(first_func)
    20200    0.115    0.000    0.115    0.000 main.py:161(my_utility)
       20    0.000    0.000    0.001    0.000 main.py:172(second_func)
```

The my_utility function is clearly the source of most execution time, but it's not immediately obvious why that function is called so much. If you search through the program's code, you'll find multiple call sites for my_utility and still be confused.

To deal with this, the Python profiler provides the print_callers method to show which callers contributed to the profiling information of each function:

```
stats.print_callers()
```

This profiler statistics table shows functions called on the left and which function was responsible for making the call on the right. Here, it's clear that my_utility is most used by first_func:

```
>>>
   Ordered by: cumulative time

Function                       was called by...
                                 ncalls  tottime  cumtime
main.py:176(my_program)        <-
main.py:168(first_func)        <-     20    0.003    0.117  main.py:176(my_program)
main.py:161(my_utility)        <-  20000    0.114    0.114  main.py:168(first_func)
                                     200    0.001    0.001  main.py:172(second_func)
Profiling.md:172(second_func)  <-     20    0.000    0.001  main.py:176(my_program)
```

Things to Remember

✦ It's important to profile Python programs before optimizing because the sources of slowdowns are often obscure.

✦ Use the cProfile module instead of the profile module because it provides more accurate profiling information.

✦ The Profile object's runcall method provides everything you need to profile a tree of function calls in isolation.

✦ The Stats object lets you select and print the subset of profiling information you need to see to understand your program's performance.

Item 71: Prefer deque for Producer–Consumer Queues

A common need in writing programs is a first-in, first-out (FIFO) queue, which is also known as a producer–consumer queue. A FIFO queue is used when one function gathers values to process and another function handles them in the order in which they were received. Often, programmers use Python's built-in list type as a FIFO queue.

For example, say that I have a program that's processing incoming emails for long-term archival, and it's using a list for a producer–consumer queue. Here, I define a class to represent the messages:

```
class Email:
    def __init__(self, sender, receiver, message):
        self.sender = sender
        self.receiver = receiver
        self.message = message
    ...
```

I also define a placeholder function for receiving a single email, presumably from a socket, the file system, or some other type of I/O system. The implementation of this function doesn't matter; what's important is its interface: It will either return an Email instance or raise a NoEmailError exception:

```
class NoEmailError(Exception):
    pass


def try_receive_email():
    # Returns an Email instance or raises NoEmailError
    ...
```

The producing function receives emails and enqueues them to be consumed at a later time. This function uses the append method on the list to add new messages to the end of the queue so they are processed after all messages that were previously received:

```
def produce_emails(queue):
    while True:
        try:
            email = try_receive_email()
        except NoEmailError:
            return
        else:
            queue.append(email)  # Producer
```

The consuming function does something useful with the emails. This function calls pop(0) on the queue, which removes the very first item from the list and returns it to the caller. By always processing items from the beginning of the queue, the consumer ensures that the items are processed in the order in which they were received:

```
def consume_one_email(queue):
    if not queue:
        return
    email = queue.pop(0)  # Consumer
```

```
    # Index the message for long-term archival
    ...
```

Finally, I need a looping function that connects the pieces together. This function alternates between producing and consuming until the keep_running function returns False (see Item 60: "Achieve Highly Concurrent I/O with Coroutines" on how to do this concurrently):

```
def loop(queue, keep_running):
    while keep_running():
        produce_emails(queue)
        consume_one_email(queue)

def my_end_func():
    ...

loop([], my_end_func)
```

Why not process each Email message in produce_emails as it's returned by try_receive_email? It comes down to the trade-off between latency and throughput. When using producer–consumer queues, you often want to minimize the latency of accepting new items so they can be collected as fast as possible. The consumer can then process through the backlog of items at a consistent pace—one item per loop in this case—which provides a stable performance profile and consistent throughput at the cost of end-to-end latency (see Item 55: "Use Queue to Coordinate Work Between Threads" for related best practices).

Using a list for a producer–consumer queue like this works fine up to a point, but as the *cardinality*—the number of items in the list—increases, the list type's performance can degrade superlinearly. To analyze the performance of using list as a FIFO queue, I can run some micro-benchmarks using the timeit built-in module. Here, I define a benchmark for the performance of adding new items to the queue using the append method of list (matching the producer function's usage):

```
import timeit

def print_results(count, tests):
    avg_iteration = sum(tests) / len(tests)
    print(f'Count {count:>5,} takes {avg_iteration:.6f}s')
    return count, avg_iteration

def list_append_benchmark(count):
    def run(queue):
```

```
        for i in range(count):
            queue.append(i)

    tests = timeit.repeat(
        setup='queue = []',
        stmt='run(queue)',
        globals=locals(),
        repeat=1000,
        number=1)

    return print_results(count, tests)
```

Running this benchmark function with different levels of cardinality lets me compare its performance in relationship to data size:

```
def print_delta(before, after):
    before_count, before_time = before
    after_count, after_time = after
    growth = 1 + (after_count - before_count) / before_count
    slowdown = 1 + (after_time - before_time) / before_time
    print(f'{growth:>4.1f}x data size, {slowdown:>4.1f}x time')

baseline = list_append_benchmark(500)
for count in (1_000, 2_000, 3_000, 4_000, 5_000):
    comparison = list_append_benchmark(count)
    print_delta(baseline, comparison)

>>>
Count   500 takes 0.000039s

Count 1,000 takes 0.000073s
 2.0x data size,  1.9x time

Count 2,000 takes 0.000121s
 4.0x data size,  3.1x time

Count 3,000 takes 0.000172s
 6.0x data size,  4.5x time

Count 4,000 takes 0.000240s
 8.0x data size,  6.2x time

Count 5,000 takes 0.000304s
10.0x data size,  7.9x time
```

This shows that the append method takes roughly constant time for the list type, and the total time for enqueueing scales linearly as the data size increases. There is overhead for the list type to increase its capacity under the covers as new items are added, but it's reasonably low and is amortized across repeated calls to append.

Here, I define a similar benchmark for the pop(0) call that removes items from the beginning of the queue (matching the consumer function's usage):

```
def list_pop_benchmark(count):
    def prepare():
        return list(range(count))

    def run(queue):
        while queue:
            queue.pop(0)

    tests = timeit.repeat(
        setup='queue = prepare()',
        stmt='run(queue)',
        globals=locals(),
        repeat=1000,
        number=1)

    return print_results(count, tests)
```

I can similarly run this benchmark for queues of different sizes to see how performance is affected by cardinality:

```
baseline = list_pop_benchmark(500)
for count in (1_000, 2_000, 3_000, 4_000, 5_000):
    comparison = list_pop_benchmark(count)
    print_delta(baseline, comparison)

>>>
Count    500 takes 0.000050s

Count 1,000 takes 0.000133s
 2.0x data size,  2.7x time

Count 2,000 takes 0.000347s
 4.0x data size,  6.9x time

Count 3,000 takes 0.000663s
 6.0x data size, 13.2x time
```

```
Count 4,000 takes 0.000943s
 8.0x data size, 18.8x time

Count 5,000 takes 0.001481s
10.0x data size, 29.5x time
```

Surprisingly, this shows that the total time for dequeuing items from a list with pop(0) scales quadratically as the length of the queue increases. The cause is that pop(0) needs to move every item in the list back an index, effectively reassigning the entire list's contents. I need to call pop(0) for every item in the list, and thus I end up doing roughly len(queue) * len(queue) operations to consume the queue. This doesn't scale.

Python provides the deque class from the collections built-in module to solve this problem. deque is a *double-ended queue* implementation. It provides constant time operations for inserting or removing items from its beginning or end. This makes it ideal for FIFO queues.

To use the deque class, the call to append in produce_emails can stay the same as it was when using a list for the queue. The list.pop method call in consume_one_email must change to call the deque.popleft method with no arguments instead. And the loop method must be called with a deque instance instead of a list. Everything else stays the same. Here, I redefine the one function affected to use the new method and run loop again:

```
import collections

def consume_one_email(queue):
    if not queue:
        return
    email = queue.popleft()  # Consumer
    # Process the email message
    ...

def my_end_func():
    ...

loop(collections.deque(), my_end_func)
```

I can run another version of the benchmark to verify that append performance (matching the producer function's usage) has stayed roughly the same (modulo a constant factor):

```
def deque_append_benchmark(count):
    def prepare():
        return collections.deque()
```

```
    def run(queue):
        for i in range(count):
            queue.append(i)

    tests = timeit.repeat(
        setup='queue = prepare()',
        stmt='run(queue)',
        globals=locals(),
        repeat=1000,
        number=1)
    return print_results(count, tests)

baseline = deque_append_benchmark(500)
for count in (1_000, 2_000, 3_000, 4_000, 5_000):
    comparison = deque_append_benchmark(count)
    print_delta(baseline, comparison)

>>>
Count    500 takes 0.000029s

Count 1,000 takes 0.000059s
 2.0x data size,  2.1x time

Count 2,000 takes 0.000121s
 4.0x data size,  4.2x time

Count 3,000 takes 0.000171s
 6.0x data size,  6.0x time

Count 4,000 takes 0.000243s
 8.0x data size,  8.5x time

Count 5,000 takes 0.000295s
10.0x data size, 10.3x time
```

And I can benchmark the performance of calling popleft to mimic the consumer function's usage of deque:

```
def dequeue_popleft_benchmark(count):
    def prepare():
        return collections.deque(range(count))

    def run(queue):
        while queue:
            queue.popleft()

    tests = timeit.repeat(
```

```
        setup='queue = prepare()',
        stmt='run(queue)',
        globals=locals(),
        repeat=1000,
        number=1)

    return print_results(count, tests)

baseline = dequeue_popleft_benchmark(500)
for count in (1_000, 2_000, 3_000, 4_000, 5_000):
    comparison = dequeue_popleft_benchmark(count)
    print_delta(baseline, comparison)

>>>
Count    500 takes 0.000024s

Count 1,000 takes 0.000050s
 2.0x data size,  2.1x time

Count 2,000 takes 0.000100s
 4.0x data size,  4.2x time

Count 3,000 takes 0.000152s
 6.0x data size,  6.3x time

Count 4,000 takes 0.000207s
 8.0x data size,  8.6x time

Count 5,000 takes 0.000265s
10.0x data size, 11.0x time
```

The popleft usage scales linearly instead of displaying the super-linear behavior of pop(0) that I measured before—hooray! If you know that the performance of a program critically depends on the speed of producer–consumer queues, then deque is a great choice. If you're not sure, then you should instrument your program to find out (see Item 70: "Profile Before Optimizing").

Things to Remember

✦ The list type can be used as a FIFO queue by having the producer call append to add items and the consumer call pop(0) to receive items. However, this may cause problems because the performance of pop(0) degrades superlinearly as the queue length increases.

✦ The deque class from the collections built-in module takes constant time—regardless of length—for append and popleft, making it ideal for FIFO queues.

Item 72: Consider Searching Sorted Sequences with bisect

It's common to find yourself with a large amount of data in memory as a sorted list that you then want to search. For example, you may have loaded an English language dictionary to use for spell checking, or perhaps a list of dated financial transactions to audit for correctness.

Regardless of the data your specific program needs to process, searching for a specific value in a list takes linear time proportional to the list's length when you call the index method:

```
data = list(range(10**5))
index = data.index(91234)
assert index == 91234
```

If you're not sure whether the exact value you're searching for is in the list, then you may want to search for the closest index that is equal to or exceeds your goal value. The simplest way to do this is to linearly scan the list and compare each item to your goal value:

```
def find_closest(sequence, goal):
    for index, value in enumerate(sequence):
        if goal < value:
            return index
    raise ValueError(f'{goal} is out of bounds')

index = find_closest(data, 91234.56)
assert index == 91235
```

Python's built-in bisect module provides better ways to accomplish these types of searches through ordered lists. You can use the bisect_left function to do an efficient binary search through any sequence of sorted items. The index it returns will either be where the item is already present in the list or where you'd want to insert the item in the list to keep it in sorted order:

```
from bisect import bisect_left

index = bisect_left(data, 91234)      # Exact match
assert index == 91234
```

```
index = bisect_left(data, 91234.56)  # Closest match
assert index == 91235
```

The complexity of the binary search algorithm used by the bisect module is logarithmic. This means searching in a list of length 1 million takes roughly the same amount of time with bisect as linearly searching a list of length 20 using the list.index method (math.log2(10**6) == 19.93...). It's way faster!

I can verify this speed improvement for the example from above by using the timeit built-in module to run a micro-benchmark:

```
import random
import timeit

size = 10**5
iterations = 1000

data = list(range(size))
to_lookup = [random.randint(0, size)
             for _ in range(iterations)]

def run_linear(data, to_lookup):
    for index in to_lookup:
        data.index(index)

def run_bisect(data, to_lookup):
    for index in to_lookup:
        bisect_left(data, index)

baseline = timeit.timeit(
    stmt='run_linear(data, to_lookup)',
    globals=globals(),
    number=10)
print(f'Linear search takes {baseline:.6f}s')

comparison = timeit.timeit(
    stmt='run_bisect(data, to_lookup)',
    globals=globals(),
    number=10)
print(f'Bisect search takes {comparison:.6f}s')

slowdown = 1 + ((baseline - comparison) / comparison)
print(f'{slowdown:.1f}x time')
```

```
>>>
Linear search takes 5.370117s
Bisect search takes 0.005220s
1028.7x time
```

The best part about bisect is that it's not limited to the list type; you can use it with any Python object that acts like a sequence (see Item 43: "Inherit from collections.abc for Custom Container Types" for how to do that). The module also provides additional features for more advanced situations (see help(bisect)).

Things to Remember

✦ Searching sorted data contained in a list takes linear time using the index method or a for loop with simple comparisons.

✦ The bisect built-in module's bisect_left function takes logarithmic time to search for values in sorted lists, which can be orders of magnitude faster than other approaches.

Item 73: Know How to Use heapq for Priority Queues

One of the limitations of Python's other queue implementations (see Item 71: "Prefer deque for Producer–Consumer Queues" and Item 55: "Use Queue to Coordinate Work Between Threads") is that they are first-in, first-out (FIFO) queues: Their contents are sorted by the order in which they were received. Often, you need a program to process items in order of relative importance instead. To accomplish this, a *priority queue* is the right tool for the job.

For example, say that I'm writing a program to manage books borrowed from a library. There are people constantly borrowing new books. There are people returning their borrowed books on time. And there are people who need to be reminded to return their overdue books. Here, I define a class to represent a book that's been borrowed:

```
class Book:
    def __init__(self, title, due_date):
        self.title = title
        self.due_date = due_date
```

I need a system that will send reminder messages when each book passes its due date. Unfortunately, I can't use a FIFO queue for this because the amount of time each book is allowed to be borrowed varies based on its recency, popularity, and other factors. For example, a book that is borrowed today may be due back later than a book that's borrowed tomorrow. Here, I achieve this behavior by using a standard list and sorting it by due_date each time a new Book is added:

```
def add_book(queue, book):
    queue.append(book)
    queue.sort(key=lambda x: x.due_date, reverse=True)

queue = []
add_book(queue, Book('Don Quixote', '2019-06-07'))
add_book(queue, Book('Frankenstein', '2019-06-05'))
add_book(queue, Book('Les Misérables', '2019-06-08'))
add_book(queue, Book('War and Peace', '2019-06-03'))
```

If I can assume that the queue of borrowed books is always in sorted order, then all I need to do to check for overdue books is to inspect the final element in the list. Here, I define a function to return the next overdue book, if any, and remove it from the queue:

```
class NoOverdueBooks(Exception):
    pass

def next_overdue_book(queue, now):
    if queue:
        book = queue[-1]
        if book.due_date < now:
            queue.pop()
            return book

    raise NoOverdueBooks
```

I can call this function repeatedly to get overdue books to remind people about in the order of most overdue to least overdue:

```
now = '2019-06-10'

found = next_overdue_book(queue, now)
print(found.title)

found = next_overdue_book(queue, now)
print(found.title)

>>>
War and Peace
Frankenstein
```

If a book is returned before the due date, I can remove the scheduled reminder message by removing the Book from the list:

```
def return_book(queue, book):
    queue.remove(book)

queue = []
book = Book('Treasure Island', '2019-06-04')

add_book(queue, book)
print('Before return:', [x.title for x in queue])

return_book(queue, book)
print('After return: ', [x.title for x in queue])

>>>
Before return: ['Treasure Island']
After return:  []
```

And I can confirm that when all books are returned, the return_book function will raise the right exception (see Item 20: "Prefer Raising Exceptions to Returning None"):

```
try:
    next_overdue_book(queue, now)
except NoOverdueBooks:
    pass            # Expected
else:
    assert False  # Doesn't happen
```

However, the computational complexity of this solution isn't ideal. Although checking for and removing an overdue book has a constant cost, every time I add a book, I pay the cost of sorting the whole list again. If I have len(queue) books to add, and the cost of sorting them is roughly len(queue) * math.log(len(queue)), the time it takes to add books will grow superlinearly (len(queue) * len(queue) * math.log(len(queue))).

Here, I define a micro-benchmark to measure this performance behavior experimentally by using the timeit built-in module (see Item 71: "Prefer deque for Producer–Consumer Queues" for the implementation of print_results and print_delta):

```
import random
import timeit

def print_results(count, tests):
    ...
```

```
def print_delta(before, after):
    ...


def list_overdue_benchmark(count):
    def prepare():
        to_add = list(range(count))
        random.shuffle(to_add)
        return [], to_add

    def run(queue, to_add):
        for i in to_add:
            queue.append(i)
            queue.sort(reverse=True)

        while queue:
            queue.pop()

    tests = timeit.repeat(
        setup='queue, to_add = prepare()',
        stmt=f'run(queue, to_add)',
        globals=locals(),
        repeat=100,
        number=1)

    return print_results(count, tests)
```

I can verify that the runtime of adding and removing books from the
queue scales superlinearly as the number of books being borrowed
increases:

```
baseline = list_overdue_benchmark(500)
for count in (1_000, 1_500, 2_000):
    comparison = list_overdue_benchmark(count)
    print_delta(baseline, comparison)
```

```
>>>
Count    500 takes 0.001138s

Count 1,000 takes 0.003317s
 2.0x data size,  2.9x time

Count 1,500 takes 0.007744s
 3.0x data size,  6.8x time

Count 2,000 takes 0.014739s
 4.0x data size, 13.0x time
```

When a book is returned before the due date, I need to do a linear scan in order to find the book in the queue and remove it. Removing a book causes all subsequent items in the `list` to be shifted back an index, which has a high cost that also scales superlinearly. Here, I define another micro-benchmark to test the performance of returning a book using this function:

```
def list_return_benchmark(count):
    def prepare():
        queue = list(range(count))
        random.shuffle(queue)

        to_return = list(range(count))
        random.shuffle(to_return)

        return queue, to_return

    def run(queue, to_return):
        for i in to_return:
            queue.remove(i)

    tests = timeit.repeat(
        setup='queue, to_return = prepare()',
        stmt=f'run(queue, to_return)',
        globals=locals(),
        repeat=100,
        number=1)

    return print_results(count, tests)
```

And again, I can verify that indeed the performance degrades super-linearly as the number of books increases:

```
baseline = list_return_benchmark(500)
for count in (1_000, 1_500, 2_000):
    comparison = list_return_benchmark(count)
    print_delta(baseline, comparison)

>>>
Count    500 takes 0.000898s

Count 1,000 takes 0.003331s
 2.0x data size,   3.7x time

Count 1,500 takes 0.007674s
 3.0x data size,   8.5x time
```

```
Count 2,000 takes 0.013721s
 4.0x data size, 15.3x time
```

Using the methods of list may work for a tiny library, but it certainly won't scale to the size of the Great Library of Alexandria, as I want it to!

Fortunately, Python has the built-in heapq module that solves this problem by implementing priority queues efficiently. A *heap* is a data structure that allows for a list of items to be maintained where the computational complexity of adding a new item or removing the smallest item has logarithmic computational complexity (i.e., even better than linear scaling). In this library example, smallest means the book with the earliest due date. The best part about this module is that you don't have to understand how heaps are implemented in order to use its functions correctly.

Here, I reimplement the add_book function using the heapq module. The queue is still a plain list. The heappush function replaces the list.append call from before. And I no longer have to call list.sort on the queue:

```
from heapq import heappush


def add_book(queue, book):
    heappush(queue, book)
```

If I try to use this with the Book class as previously defined, I get this somewhat cryptic error:

```
queue = []
add_book(queue, Book('Little Women', '2019-06-05'))
add_book(queue, Book('The Time Machine', '2019-05-30'))

>>>
Traceback ...
TypeError: '<' not supported between instances of 'Book' and
➥'Book'
```

The heapq module requires items in the priority queue to be comparable and have a natural sort order (see Item 14: "Sort by Complex Criteria Using the key Parameter" for details). You can quickly give the Book class this behavior by using the total_ordering class decorator from the functools built-in module (see Item 51: "Prefer Class Decorators Over Metaclasses for Composable Class Extensions" for background) and implementing the __lt__ special method (see Item 43: "Inherit from collections.abc for Custom Container Types" for background). Here, I redefine the class with a less-than method that simply compares the due_date fields between two Book instances:

```
import functools

@functools.total_ordering
class Book:
    def __init__(self, title, due_date):
        self.title = title
        self.due_date = due_date

    def __lt__(self, other):
        return self.due_date < other.due_date
```

Now, I can add books to the priority queue by using the heapq.heappush function without issues:

```
queue = []
add_book(queue, Book('Pride and Prejudice', '2019-06-01'))
add_book(queue, Book('The Time Machine', '2019-05-30'))
add_book(queue, Book('Crime and Punishment', '2019-06-06'))
add_book(queue, Book('Wuthering Heights', '2019-06-12'))
```

Alternatively, I can create a list with all of the books in any order and then use the sort method of list to produce the heap:

```
queue = [
    Book('Pride and Prejudice', '2019-06-01'),
    Book('The Time Machine', '2019-05-30'),
    Book('Crime and Punishment', '2019-06-06'),
    Book('Wuthering Heights', '2019-06-12'),
]
queue.sort()
```

Or I can use the heapq.heapify function to create a heap in linear time (as opposed to the sort method's len(queue) * log(len(queue)) complexity):

```
from heapq import heapify

queue = [
    Book('Pride and Prejudice', '2019-06-01'),
    Book('The Time Machine', '2019-05-30'),
    Book('Crime and Punishment', '2019-06-06'),
    Book('Wuthering Heights', '2019-06-12'),
]
heapify(queue)
```

To check for overdue books, I inspect the first element in the list instead of the last, and then I use the heapq.heappop function instead of the list.pop function:

```
from heapq import heappop

def next_overdue_book(queue, now):
    if queue:
        book = queue[0]           # Most overdue first
        if book.due_date < now:
            heappop(queue)        # Remove the overdue book
            return book

    raise NoOverdueBooks
```

Now, I can find and remove overdue books in order until there are none left for the current time:

```
now = '2019-06-02'

book = next_overdue_book(queue, now)
print(book.title)

book = next_overdue_book(queue, now)
print(book.title)

try:
    next_overdue_book(queue, now)
except NoOverdueBooks:
    pass            # Expected
else:
    assert False  # Doesn't happen
>>>
The Time Machine
Pride and Prejudice
```

I can write another micro-benchmark to test the performance of this implementation that uses the heapq module:

```
def heap_overdue_benchmark(count):
    def prepare():
        to_add = list(range(count))
        random.shuffle(to_add)
        return [], to_add

    def run(queue, to_add):
        for i in to_add:
```

```
            heappush(queue, i)
        while queue:
            heappop(queue)

    tests = timeit.repeat(
        setup='queue, to_add = prepare()',
        stmt=f'run(queue, to_add)',
        globals=locals(),
        repeat=100,
        number=1)

    return print_results(count, tests)
```

This benchmark experimentally verifies that the heap-based priority queue implementation scales much better (roughly len(queue) * math.log(len(queue))), without superlinearly degrading performance:

```
baseline = heap_overdue_benchmark(500)
for count in (1_000, 1_500, 2_000):
    comparison = heap_overdue_benchmark(count)
    print_delta(baseline, comparison)

>>>
Count   500 takes 0.000150s

Count 1,000 takes 0.000325s
 2.0x data size,  2.2x time

Count 1,500 takes 0.000528s
 3.0x data size,  3.5x time

Count 2,000 takes 0.000658s
 4.0x data size,  4.4x time
```

With the heapq implementation, one question remains: How should I handle returns that are on time? The solution is to never remove a book from the priority queue until its due date. At that time, it will be the first item in the list, and I can simply ignore the book if it's already been returned. Here, I implement this behavior by adding a new field to track the book's return status:

```
@functools.total_ordering
class Book:
    def __init__(self, title, due_date):
        self.title = title
        self.due_date = due_date
```

```
      self.returned = False  # New field
```

   ...

Then, I change the next_overdue_book function to repeatedly ignore any book that's already been returned:

```
def next_overdue_book(queue, now):
    while queue:
        book = queue[0]
        if book.returned:
            heappop(queue)
            continue

        if book.due_date < now:
            heappop(queue)
            return book

        break

    raise NoOverdueBooks
```

This approach makes the return_book function extremely fast because it makes no modifications to the priority queue:

```
def return_book(queue, book):
    book.returned = True
```

The downside of this solution for returns is that the priority queue may grow to the maximum size it would have needed if all books from the library were checked out and went overdue. Although the queue operations will be fast thanks to heapq, this storage overhead may take significant memory (see Item 81: "Use tracemalloc to Understand Memory Usage and Leaks" for how to debug such usage).

That said, if you're trying to build a robust system, you need to plan for the worst-case scenario; thus, you should expect that it's possible for every library book to go overdue for some reason (e.g., a natural disaster closes the road to the library). This memory cost is a design consideration that you should have already planned for and mitigated through additional constraints (e.g., imposing a maximum number of simultaneously lent books).

Beyond the priority queue primitives that I've used in this example, the heapq module provides additional functionality for advanced use cases (see help(heapq)). The module is a great choice when its functionality matches the problem you're facing (see the queue.PriorityQueue class for another thread-safe option).

Things to Remember

✦ Priority queues allow you to process items in order of importance instead of in first-in, first-out order.

✦ If you try to use `list` operations to implement a priority queue, your program's performance will degrade superlinearly as the queue grows.

✦ The `heapq` built-in module provides all of the functions you need to implement a priority queue that scales efficiently.

✦ To use `heapq`, the items being prioritized must have a natural sort order, which requires special methods like `__lt__` to be defined for classes.

Item 74: Consider `memoryview` and `bytearray` for Zero-Copy Interactions with `bytes`

Although Python isn't able to parallelize CPU-bound computation without extra effort (see Item 64: "Consider concurrent.futures for True Parallelism"), it is able to support high-throughput, parallel I/O in a variety of ways (see Item 53: "Use Threads for Blocking I/O, Avoid for Parallelism" and Item 60: "Achieve Highly Concurrent I/O with Coroutines"). That said, it's surprisingly easy to use these I/O tools the wrong way and reach the conclusion that the language is too slow for even I/O-bound workloads.

For example, say that I'm building a media server to stream television or movies over a network to users so they can watch without having to download the video data in advance. One of the key features of such a system is the ability for users to move forward or backward in the video playback so they can skip or repeat parts. In the client program, I can implement this by requesting a chunk of data from the server corresponding to the new time index selected by the user:

```
def timecode_to_index(video_id, timecode):
    ...
    # Returns the byte offset in the video data

def request_chunk(video_id, byte_offset, size):
    ...
    # Returns size bytes of video_id's data from the offset

video_id = ...
timecode = '01:09:14:28'
byte_offset = timecode_to_index(video_id, timecode)
```

```
size = 20 * 1024 * 1024
video_data = request_chunk(video_id, byte_offset, size)
```

How would you implement the server-side handler that receives the request_chunk request and returns the corresponding 20 MB chunk of video data? For the sake of this example, I assume that the command and control parts of the server have already been hooked up (see Item 61: "Know How to Port Threaded I/O to asyncio" for what that requires). I focus here on the last steps where the requested chunk is extracted from gigabytes of video data that's cached in memory and is then sent over a socket back to the client. Here's what the implementation would look like:

```
socket = ...              # socket connection to client
video_data = ...          # bytes containing data for video_id
byte_offset = ...         # Requested starting position
size = 20 * 1024 * 1024   # Requested chunk size

chunk = video_data[byte_offset:byte_offset + size]
socket.send(chunk)
```

The latency and throughput of this code will come down to two factors: how much time it takes to slice the 20 MB video chunk from video_data, and how much time the socket takes to transmit that data to the client. If I assume that the socket is infinitely fast, I can run a micro-benchmark by using the timeit built-in module to understand the performance characteristics of slicing bytes instances this way to create chunks (see Item 11: "Know How to Slice Sequences" for background):

```
import timeit

def run_test():
    chunk = video_data[byte_offset:byte_offset + size]
    # Call socket.send(chunk), but ignoring for benchmark

result = timeit.timeit(
    stmt='run_test()',
    globals=globals(),
    number=100) / 100

print(f'{result:0.9f} seconds')

>>>
0.004925669 seconds
```

It took roughly 5 milliseconds to extract the 20 MB slice of data to transmit to the client. That means the overall throughput of my server is limited to a theoretical maximum of 20 MB / 5 milliseconds = 7.3 GB / second, since that's the fastest I can extract the video data from memory. My server will also be limited to 1 CPU-second / 5 milliseconds = 200 clients requesting new chunks in parallel, which is tiny compared to the tens of thousands of simultaneous connections that tools like the `asyncio` built-in module can support. The problem is that slicing a `bytes` instance causes the underlying data to be copied, which takes CPU time.

A better way to write this code is by using Python's built-in `memoryview` type, which exposes CPython's high-performance *buffer protocol* to programs. The buffer protocol is a low-level C API that allows the Python runtime and C extensions to access the underlying data buffers that are behind objects like `bytes` instances. The best part about `memoryview` instances is that slicing them results in another `memoryview` instance without copying the underlying data. Here, I create a `memoryview` wrapping a `bytes` instance and inspect a slice of it:

```
data = b'shave and a haircut, two bits'
view = memoryview(data)
chunk = view[12:19]
print(chunk)
print('Size:            ', chunk.nbytes)
print('Data in view:    ', chunk.tobytes())
print('Underlying data:', chunk.obj)

>>>
<memory at 0x10951fb80>
Size:             7
Data in view:     b'haircut'
Underlying data: b'shave and a haircut, two bits'
```

By enabling *zero-copy* operations, `memoryview` can provide enormous speedups for code that needs to quickly process large amounts of memory, such as numerical C extensions like NumPy and I/O-bound programs like this one. Here, I replace the simple `bytes` slicing from above with `memoryview` slicing instead and repeat the same micro-benchmark:

```
video_view = memoryview(video_data)

def run_test():
    chunk = video_view[byte_offset:byte_offset + size]
    # Call socket.send(chunk), but ignoring for benchmark
```

```
result = timeit.timeit(
    stmt='run_test()',
    globals=globals(),
    number=100) / 100

print(f'{result:0.9f} seconds')

>>>
0.000000250 seconds
```

The result is 250 nanoseconds. Now the theoretical maximum through-put of my server is 20 MB / 250 nanoseconds = 164 TB / second. For parallel clients, I can theoretically support up to 1 CPU-second / 250 nanoseconds = 4 million. That's more like it! This means that now my program is entirely bound by the underlying performance of the socket connection to the client, not by CPU constraints.

Now, imagine that the data must flow in the other direction, where some clients are sending live video streams to the server in order to broadcast them to other users. In order to do this, I need to store the latest video data from the user in a cache that other clients can read from. Here's what the implementation of reading 1 MB of new data from the incoming client would look like:

```
socket = ...        # socket connection to the client
video_cache = ...   # Cache of incoming video stream
byte_offset = ...   # Incoming buffer position
size = 1024 * 1024  # Incoming chunk size

chunk = socket.recv(size)
video_view = memoryview(video_cache)
before = video_view[:byte_offset]
after = video_view[byte_offset + size:]
new_cache = b''.join([before, chunk, after])
```

The `socket.recv` method returns a `bytes` instance. I can splice the new data with the existing cache at the current `byte_offset` by using simple slicing operations and the `bytes.join` method. To understand the performance of this, I can run another micro-benchmark. I'm using a dummy socket, so the performance test is only for the memory operations, not the I/O interaction:

```
def run_test():
    chunk = socket.recv(size)
    before = video_view[:byte_offset]
    after = video_view[byte_offset + size:]
    new_cache = b''.join([before, chunk, after])
```

```
result = timeit.timeit(
    stmt='run_test()',
    globals=globals(),
    number=100) / 100

print(f'{result:0.9f} seconds')

>>>
0.033520550 seconds
```

It takes 33 milliseconds to receive 1 MB and update the video cache. This means my maximum receive throughput is 1 MB / 33 milliseconds = 31 MB / second, and I'm limited to 31 MB / 1 MB = 31 simultaneous clients streaming in video data this way. This doesn't scale.

A better way to write this code is to use Python's built-in bytearray type in conjunction with memoryview. One limitation with bytes instances is that they are read-only and don't allow for individual indexes to be updated:

```
my_bytes = b'hello'
my_bytes[0] = b'\x79'

>>>
Traceback ...
TypeError: 'bytes' object does not support item assignment
```

The bytearray type is like a mutable version of bytes that allows for arbitrary positions to be overwritten. bytearray uses integers for its values instead of bytes:

```
my_array = bytearray(b'hello')
my_array[0] = 0x79
print(my_array)

>>>
bytearray(b'yello')
```

A memoryview can also be used to wrap a bytearray. When you slice such a memoryview, the resulting object can be used to assign data to a particular portion of the underlying buffer. This eliminates the copying costs from above that were required to splice the bytes instances back together after data was received from the client:

```
my_array = bytearray(b'row, row, row your boat')
my_view = memoryview(my_array)
write_view = my_view[3:13]
write_view[:] = b'-10 bytes-'
print(my_array)
```

```
>>>
bytearray(b'row-10 bytes- your boat')
```

Many library methods in Python, such as `socket.recv_into` and `RawIOBase.readinto`, use the buffer protocol to receive or read data quickly. The benefit of these methods is that they avoid allocating memory and creating another copy of the data; what's received goes straight into an existing buffer. Here, I use `socket.recv_into` along with a `memoryview` slice to receive data into an underlying `bytearray` without the need for splicing:

```
video_array = bytearray(video_cache)
write_view = memoryview(video_array)
chunk = write_view[byte_offset:byte_offset + size]
socket.recv_into(chunk)
```

I can run another micro-benchmark to compare the performance of this approach to the earlier example that used `socket.recv`:

```
def run_test():
    chunk = write_view[byte_offset:byte_offset + size]
    socket.recv_into(chunk)

result = timeit.timeit(
    stmt='run_test()',
    globals=globals(),
    number=100) / 100

print(f'{result:0.9f} seconds')

>>>
0.000033925 seconds
```

It took 33 microseconds to receive a 1 MB video transmission. This means my server can support 1 MB / 33 microseconds = 31 GB / second of max throughput, and 31 GB / 1 MB = 31,000 parallel streaming clients. That's the type of scalability that I'm looking for!

Things to Remember

✦ The `memoryview` built-in type provides a zero-copy interface for reading and writing slices of objects that support Python's high-performance buffer protocol.

✦ The `bytearray` built-in type provides a mutable bytes-like type that can be used for zero-copy data reads with functions like `socket.recv_from`.

✦ A `memoryview` can wrap a `bytearray`, allowing for received data to be spliced into an arbitrary buffer location without copying costs.

Chapter 9

Testing and Debugging

Python doesn't have compile-time static type checking. There's nothing in the interpreter that will ensure that your program will work correctly when you run it. Python does support optional type annotations that can be used in static analysis to detect many kinds of bugs (see Item 90: "Consider Static Analysis via typing to Obviate Bugs" for details). However, it's still fundamentally a dynamic language, and anything is possible. With Python, you ultimately don't know if the functions your program calls will be defined at runtime, even when their existence is evident in the source code. This dynamic behavior is both a blessing and a curse.

The large numbers of Python programmers out there say it's worth going without compile-time static type checking because of the productivity gained from the resulting brevity and simplicity. But most people using Python have at least one horror story about a program encountering a boneheaded error at runtime. One of the worst examples I've heard of involved a SyntaxError being raised in production as a side effect of a dynamic import (see Item 88: "Know How to Break Circular Dependencies"), resulting in a crashed server process. The programmer I know who was hit by this surprising occurrence has since ruled out using Python ever again.

But I have to wonder, why wasn't the code more well tested before the program was deployed to production? Compile-time static type safety isn't everything. You should always test your code, regardless of what language it's written in. However, I'll admit that in Python it may be more important to write tests to verify correctness than in other languages. Luckily, the same dynamic features that create risks also make it extremely easy to write tests for your code and to debug malfunctioning programs. You can use Python's dynamic nature and easily overridable behaviors to implement tests and ensure that your programs work as expected.

You should think of tests as an insurance policy on your code. Good tests give you confidence that your code is correct. If you refactor or expand your code, tests that verify behavior—*not* implementation— make it easy to identify what's changed. It sounds counterintuitive, but having good tests actually makes it easier to modify Python code, not harder.

Item 75: Use repr Strings for Debugging Output

When debugging a Python program, the print function and format strings (see Item 4: "Prefer Interpolated F-Strings Over C-style Format Strings and str.format"), or output via the logging built-in module, will get you surprisingly far. Python internals are often easy to access via plain attributes (see Item 42: "Prefer Public Attributes Over Private Ones"). All you need to do is call print to see how the state of your program changes while it runs and understand where it goes wrong.

The print function outputs a human-readable string version of whatever you supply it. For example, printing a basic string prints the contents of the string without the surrounding quote characters:

```
print('foo bar')
```

```
>>>
foo bar
```

This is equivalent to all of these alternatives:

- Calling the str function before passing the value to print
- Using the '%s' format string with the% operator
- Default formatting of the value with an f-string
- Calling the format built-in function
- Explicitly calling the __format__ special method
- Explicitly calling the __str__ special method

Here, I verify this behavior:

```
my_value = 'foo bar'
print(str(my_value))
print('%s' % my_value)
print(f'{my_value}')
print(format(my_value))
print(my_value.__format__('s'))
print(my_value.__str__())
```

```
>>>
foo bar
foo bar
foo bar
foo bar
foo bar
foo bar
```

The problem is that the human-readable string for a value doesn't make it clear what the actual type and its specific composition are. For example, notice how in the default output of print, you can't distinguish between the types of the number 5 and the string '5':

```
print(5)
print('5')

int_value = 5
str_value = '5'
print(f'{int_value} == {str_value} ?')
```

```
>>>
5
5
5 == 5 ?
```

If you're debugging a program with print, these type differences matter. What you almost always want while debugging is to see the repr version of an object. The repr built-in function returns the *printable representation* of an object, which should be its most clearly understandable string representation. For most built-in types, the string returned by repr is a valid Python expression:

```
a = '\x07'
print(repr(a))
```

```
>>>
'\x07'
```

Passing the value from repr to the eval built-in function should result in the same Python object that you started with (and, of course, in practice you should only use eval with extreme caution):

```
b = eval(repr(a))
assert a == b
```

When you're debugging with print, you should call repr on a value before printing to ensure that any difference in types is clear:

```
print(repr(5))
print(repr('5'))
```

```
>>>
5
'5'
```

This is equivalent to using the '%r' format string with the % operator or an f-string with the !r type conversion:

```
print('%r' % 5)
print('%r' % '5')

int_value = 5
str_value = '5'
print(f'{int_value!r} != {str_value!r}')

>>>
5
'5'
5 != '5'
```

For instances of Python classes, the default human-readable string value is the same as the repr value. This means that passing an instance to print will do the right thing, and you don't need to explicitly call repr on it. Unfortunately, the default implementation of repr for object subclasses isn't especially helpful. For example, here I define a simple class and then print one of its instances:

```
class OpaqueClass:
    def __init__(self, x, y):
        self.x = x
        self.y = y

obj = OpaqueClass(1, 'foo')
print(obj)

>>>
<__main__.OpaqueClass object at 0x10963d6d0>
```

This output can't be passed to the eval function, and it says nothing about the instance fields of the object.

There are two solutions to this problem. If you have control of the class, you can define your own __repr__ special method that returns a string containing the Python expression that re-creates the object. Here, I define that function for the class above:

```
class BetterClass:
    def __init__(self, x, y):
        self.x = x
        self.y = y
```

```
    def __repr__(self):
        return f'BetterClass({self.x!r}, {self.y!r})'
```

Now the repr value is much more useful:

```
obj = BetterClass(2, 'bar')
print(obj)
```

```
>>>
BetterClass(2, 'bar')
```

When you don't have control over the class definition, you can reach into the object's instance dictionary, which is stored in the __dict__ attribute. Here, I print out the contents of an OpaqueClass instance:

```
obj = OpaqueClass(4, 'baz')
print(obj.__dict__)
```

```
>>>
{'x': 4, 'y': 'baz'}
```

Things to Remember

✦ Calling print on built-in Python types produces the human-readable string version of a value, which hides type information.

✦ Calling repr on built-in Python types produces the printable string version of a value. These repr strings can often be passed to the eval built-in function to get back the original value.

✦ %s in format strings produces human-readable strings like str. %r produces printable strings like repr. F-strings produce human-readable strings for replacement text expressions unless you specify the !r suffix.

✦ You can define the __repr__ special method on a class to customize the printable representation of instances and provide more detailed debugging information.

Item 76: Verify Related Behaviors in TestCase Subclasses

The canonical way to write tests in Python is to use the unittest built-in module. For example, say I have the following utility function defined in utils.py that I would like to verify works correctly across a variety of inputs:

```
# utils.py
def to_str(data):
```

```
    if isinstance(data, str):
        return data
    elif isinstance(data, bytes):
        return data.decode('utf-8')
    else:
        raise TypeError('Must supply str or bytes, '
                        'found: %r' % data)
```

To define tests, I create a second file named test_utils.py or utils_test.py—the naming scheme you prefer is a style choice—that contains tests for each behavior that I expect:

```
# utils_test.py
from unittest import TestCase, main
from utils import to_str

class UtilsTestCase(TestCase):
    def test_to_str_bytes(self):
        self.assertEqual('hello', to_str(b'hello'))

    def test_to_str_str(self):
        self.assertEqual('hello', to_str('hello'))

    def test_failing(self):
        self.assertEqual('incorrect', to_str('hello'))

if __name__ == '__main__':
    main()
```

Then, I run the test file using the Python command line. In this case, two of the test methods pass and one fails, with a helpful error message about what went wrong:

```
$ python3 utils_test.py
F..
================================================================
FAIL: test_failing (__main__.UtilsTestCase)
----------------------------------------------------------------
Traceback (most recent call last):
  File "utils_test.py", line 15, in test_failing
    self.assertEqual('incorrect', to_str('hello'))
AssertionError: 'incorrect' != 'hello'
- incorrect
+ hello

----------------------------------------------------------------
```

```
Ran 3 tests in 0.002s
```

```
FAILED (failures=1)
```

Tests are organized into TestCase subclasses. Each test case is a method beginning with the word test. If a test method runs without raising any kind of Exception (including AssertionError from assert statements), the test is considered to have passed successfully. If one test fails, the TestCase subclass continues running the other test methods so you can get a full picture of how all your tests are doing instead of stopping at the first sign of trouble.

If you want to iterate quickly to fix or improve a specific test, you can run only that test method by specifying its path within the test module on the command line:

```
$ python3 utils_test.py UtilsTestCase.test_to_str_bytes
.
----------------------------------------------------------------
Ran 1 test in 0.000s
```

```
OK
```

You can also invoke the debugger from directly within test methods at specific breakpoints in order to dig more deeply into the cause of failures (see Item 80: "Consider Interactive Debugging with pdb" for how to do that).

The TestCase class provides helper methods for making assertions in your tests, such as assertEqual for verifying equality, assertTrue for verifying Boolean expressions, and many more (see help(TestCase) for the full list). These are better than the built-in assert statement because they print out all of the inputs and outputs to help you understand the exact reason the test is failing. For example, here I have the same test case written with and without using a helper assertion method:

```
# assert_test.py
from unittest import TestCase, main
from utils import to_str

class AssertTestCase(TestCase):
    def test_assert_helper(self):
        expected = 12
        found = 2 * 5
        self.assertEqual(expected, found)
```

```
    def test_assert_statement(self):
        expected = 12
        found = 2 * 5
        assert expected == found

if __name__ == '__main__':
    main()
```

Which of these failure messages seems more helpful to you?

```
$ python3 assert_test.py
FF
================================================================
FAIL: test_assert_helper (__main__.AssertTestCase)
----------------------------------------------------------------
Traceback (most recent call last):
  File "assert_test.py", line 16, in test_assert_helper
    self.assertEqual(expected, found)
AssertionError: 12 != 10


================================================================
FAIL: test_assert_statement (__main__.AssertTestCase)
----------------------------------------------------------------
Traceback (most recent call last):
  File "assert_test.py", line 11, in test_assert_statement
    assert expected == found
AssertionError


----------------------------------------------------------------
Ran 2 tests in 0.001s

FAILED (failures=2)
```

There's also an assertRaises helper method for verifying exceptions that can be used as a context manager in with statements (see Item 66: "Consider contextlib and with Statements for Reusable try/finally Behavior" for how that works). This appears similar to a try/except statement and makes it abundantly clear where the exception is expected to be raised:

```
# utils_error_test.py
from unittest import TestCase, main
from utils import to_str

class UtilsErrorTestCase(TestCase):
```

```
    def test_to_str_bad(self):
        with self.assertRaises(TypeError):
            to_str(object())

    def test_to_str_bad_encoding(self):
        with self.assertRaises(UnicodeDecodeError):
            to_str(b'\xfa\xfa')

if __name__ == '__main__':
    main()
```

You can define your own helper methods with complex logic in TestCase subclasses to make your tests more readable. Just ensure that your method names don't begin with the word test, or they'll be run as if they're test cases. In addition to calling TestCase assertion methods, these custom test helpers often use the fail method to clarify which assumption or invariant wasn't met. For example, here I define a custom test helper method for verifying the behavior of a generator:

```
# helper_test.py
from unittest import TestCase, main

def sum_squares(values):
    cumulative = 0
    for value in values:
        cumulative += value ** 2
        yield cumulative

class HelperTestCase(TestCase):
    def verify_complex_case(self, values, expected):
        expect_it = iter(expected)
        found_it = iter(sum_squares(values))
        test_it = zip(expect_it, found_it)

        for i, (expect, found) in enumerate(test_it):
            self.assertEqual(
                expect,
                found,
                f'Index {i} is wrong')

        # Verify both generators are exhausted
        try:
            next(expect_it)
        except StopIteration:
            pass
```

```
            else:
                self.fail('Expected longer than found')

            try:
                next(found_it)
            except StopIteration:
                pass
            else:
                self.fail('Found longer than expected')

    def test_wrong_lengths(self):
        values = [1.1, 2.2, 3.3]
        expected = [
            1.1**2,
        ]
        self.verify_complex_case(values, expected)

    def test_wrong_results(self):
        values = [1.1, 2.2, 3.3]
        expected = [
            1.1**2,
            1.1**2 + 2.2**2,
            1.1**2 + 2.2**2 + 3.3**2 + 4.4**2,
        ]
        self.verify_complex_case(values, expected)

if __name__ == '__main__':
    main()
```

The helper method makes the test cases short and readable, and the outputted error messages are easy to understand:

```
$ python3 helper_test.py
FF
======================================================================
FAIL: test_wrong_lengths (__main__.HelperTestCase)
----------------------------------------------------------------------
Traceback (most recent call last):
  File "helper_test.py", line 43, in test_wrong_lengths
    self.verify_complex_case(values, expected)
  File "helper_test.py", line 34, in verify_complex_case
    self.fail('Found longer than expected')
AssertionError: Found longer than expected
```

```
================================================================
FAIL: test_wrong_results (__main__.HelperTestCase)
----------------------------------------------------------------
Traceback (most recent call last):
  File "helper_test.py", line 52, in test_wrong_results
    self.verify_complex_case(values, expected)
  File "helper_test.py", line 24, in verify_complex_case
    f'Index {i} is wrong')
AssertionError: 36.3 != 16.939999999999998 : Index 2 is wrong


----------------------------------------------------------------
Ran 2 tests in 0.002s

FAILED (failures=2)
```

I usually define one TestCase subclass for each set of related tests. Sometimes, I have one TestCase subclass for each function that has many edge cases. Other times, a TestCase subclass spans all functions in a single module. I often create one TestCase subclass for testing each basic class and all of its methods.

The TestCase class also provides a subTest helper method that enables you to avoid boilerplate by defining multiple tests within a single test method. This is especially helpful for writing data-driven tests, and it allows the test method to continue testing other cases even after one of them fails (similar to the behavior of TestCase with its contained test methods). To show this, here I define an example data-driven test:

```python
# data_driven_test.py
from unittest import TestCase, main
from utils import to_str

class DataDrivenTestCase(TestCase):
    def test_good(self):
        good_cases = [
            (b'my bytes', 'my bytes'),
            ('no error', b'no error'),  # This one will fail
            ('other str', 'other str'),
            ...
        ]
        for value, expected in good_cases:
            with self.subTest(value):
                self.assertEqual(expected, to_str(value))
```

```
    def test_bad(self):
        bad_cases = [
            (object(), TypeError),
            (b'\xfa\xfa', UnicodeDecodeError),
            ...
        ]
        for value, exception in bad_cases:
            with self.subTest(value):
                with self.assertRaises(exception):
                    to_str(value)

if __name__ == '__main__':
    main()
```

The 'no error' test case fails, printing a helpful error message, but all of the other cases are still tested and confirmed to pass:

```
$ python3 data_driven_test.py
.
================================================================
FAIL: test_good (__main__.DataDrivenTestCase) [no error]
----------------------------------------------------------------
Traceback (most recent call last):
  File "testing/data_driven_test.py", line 18, in test_good
    self.assertEqual(expected, to_str(value))
AssertionError: b'no error' != 'no error'


----------------------------------------------------------------
Ran 2 tests in 0.001s

FAILED (failures=1)
```

> **Note**
> Depending on your project's complexity and testing requirements, the *pytest* open source package and its large number of community plug-ins can be especially useful.

Things to Remember

✦ You can create tests by subclassing the TestCase class from the unittest built-in module and defining one method per behavior you'd like to test. Test methods on TestCase classes must start with the word test.

✦ Use the various helper methods defined by the TestCase class, such as assertEqual, to confirm expected behaviors in your tests instead of using the built-in assert statement.

✦ Consider writing data-driven tests using the subTest helper method in order to reduce boilerplate.

Item 77: **Isolate Tests from Each Other with** setUp, tearDown, setUpModule, **and** tearDownModule

TestCase classes (see Item 76: "Verify Related Behaviors in TestCase Subclasses") often need to have the test environment set up before test methods can be run; this is sometimes called the *test harness*. To do this, you can override the setUp and tearDown methods of a TestCase subclass. These methods are called before and after each test method, respectively, so you can ensure that each test runs in isolation, which is an important best practice of proper testing.

For example, here I define a TestCase that creates a temporary directory before each test and deletes its contents after each test finishes:

```
# environment_test.py
from pathlib import Path
from tempfile import TemporaryDirectory
from unittest import TestCase, main

class EnvironmentTest(TestCase):
    def setUp(self):
        self.test_dir = TemporaryDirectory()
        self.test_path = Path(self.test_dir.name)

    def tearDown(self):
        self.test_dir.cleanup()

    def test_modify_file(self):
        with open(self.test_path / 'data.bin', 'w') as f:
            ...

if __name__ == '__main__':
    main()
```

When programs get complicated, you'll want additional tests to verify the end-to-end interactions between your modules instead of only testing code in isolation (using tools like mocks; see Item 78: "Use Mocks to Test Code with Complex Dependencies"). This is the difference between *unit tests* and *integration tests*. In Python, it's important to write both types of tests for exactly the same reason: You have no guarantee that your modules will actually work together unless you prove it.

One common problem is that setting up your test environment for integration tests can be computationally expensive and may require a lot of wall-clock time. For example, you might need to start a database process and wait for it to finish loading indexes before you can run your integration tests. This type of latency makes it impractical to do test preparation and cleanup for every test in the TestCase class's setUp and tearDown methods.

To handle this situation, the unittest module also supports module-level test harness initialization. You can configure expensive resources a single time, and then have all TestCase classes and their test methods run without repeating that initialization. Later, when all tests in the module are finished, the test harness can be torn down a single time. Here, I take advantage of this behavior by defining setUpModule and tearDownModule functions within the module containing the TestCase classes:

```
# integration_test.py
from unittest import TestCase, main

def setUpModule():
    print('* Module setup')

def tearDownModule():
    print('* Module clean-up')

class IntegrationTest(TestCase):
    def setUp(self):
        print('* Test setup')

    def tearDown(self):
        print('* Test clean-up')

    def test_end_to_end1(self):
        print('* Test 1')

    def test_end_to_end2(self):
        print('* Test 2')

if __name__ == '__main__':
    main()
```

```
$ python3 integration_test.py
* Module setup
* Test setup
* Test 1
```

```
* Test clean-up
.* Test setup
* Test 2
* Test clean-up
.* Module clean-up
```

```
----------------------------------------------------------------
```

```
Ran 2 tests in 0.000s
```

```
OK
```

I can clearly see that setUpModule is run by unittest only once, and it happens before any setUp methods are called. Similarly, tearDownModule happens after the tearDown method is called.

Things to Remember

✦ It's important to write both unit tests (for isolated functionality) and integration tests (for modules that interact with each other).

✦ Use the setUp and tearDown methods to make sure your tests are isolated from each other and have a clean test environment.

✦ For integration tests, use the setUpModule and tearDownModule module-level functions to manage any test harnesses you need for the entire lifetime of a test module and all of the TestCase classes that it contains.

Item 78: Use Mocks to Test Code with Complex Dependencies

Another common need when writing tests (see Item 76: "Verify Related Behaviors in TestCase Subclasses") is to use mocked functions and classes to simulate behaviors when it's too difficult or slow to use the real thing. For example, say that I need a program to maintain the feeding schedule for animals at the zoo. Here, I define a function to query a database for all of the animals of a certain species and return when they most recently ate:

```
class DatabaseConnection:
    ...

def get_animals(database, species):
    # Query the database
    ...
    # Return a list of (name, last_mealtime) tuples
```

How do I get a DatabaseConnection instance to use for testing this function? Here, I try to create one and pass it into the function being tested:

```
database = DatabaseConnection('localhost', '4444')

get_animals(database, 'Meerkat')

>>>
Traceback ...
DatabaseConnectionError: Not connected
```

There's no database running, so of course this fails. One solution is to actually stand up a database server and connect to it in the test. However, it's a lot of work to fully automate starting up a database, configuring its schema, populating it with data, and so on in order to just run a simple unit test. Further, it will probably take a lot of wall-clock time to set up a database server, which would slow down these unit tests and make them harder to maintain.

A better approach is to mock out the database. A *mock* lets you provide expected responses for dependent functions, given a set of expected calls. It's important not to confuse mocks with fakes. A *fake* would provide most of the behavior of the DatabaseConnection class but with a simpler implementation, such as a basic in-memory, single-threaded database with no persistence.

Python has the unittest.mock built-in module for creating mocks and using them in tests. Here, I define a Mock instance that simulates the get_animals function without actually connecting to the database:

```
from datetime import datetime
from unittest.mock import Mock

mock = Mock(spec=get_animals)
expected = [
    ('Spot', datetime(2019, 6, 5, 11, 15)),
    ('Fluffy', datetime(2019, 6, 5, 12, 30)),
    ('Jojo', datetime(2019, 6, 5, 12, 45)),
]
mock.return_value = expected
```

The Mock class creates a mock function. The return_value attribute of the mock is the value to return when it is called. The spec argument indicates that the mock should act like the given object, which is a function in this case, and error if it's used in the wrong way.

For example, here I try to treat the mock function as if it were a mock object with attributes:

```
mock.does_not_exist
```

```
>>>
Traceback ...
AttributeError: Mock object has no attribute 'does_not_exist'
```

Once it's created, I can call the mock, get its return value, and verify that what it returns matches expectations. I use a unique object value as the database argument because it won't actually be used by the mock to do anything; all I care about is that the database parameter was correctly plumbed through to any dependent functions that needed a DatabaseConnection instance in order to work (see Item 55: "Use Queue to Coordinate Work Between Threads" for another example of using sentinel object instances):

```
database = object()
result = mock(database, 'Meerkat')
assert result == expected
```

This verifies that the mock responded correctly, but how do I know if the code that called the mock provided the correct arguments? For this, the Mock class provides the assert_called_once_with method, which verifies that a single call with exactly the given parameters was made:

```
mock.assert_called_once_with(database, 'Meerkat')
```

If I supply the wrong parameters, an exception is raised, and any TestCase that the assertion is used in fails:

```
mock.assert_called_once_with(database, 'Giraffe')
```

```
>>>
Traceback ...
AssertionError: expected call not found.
Expected: mock(<object object at 0x109038790>, 'Giraffe')
Actual: mock(<object object at 0x109038790>, 'Meerkat')
```

If I actually don't care about some of the individual parameters, such as exactly which database object was used, then I can indicate that any value is okay for an argument by using the unittest.mock.ANY constant. I can also use the assert_called_with method of Mock to verify that the most recent call to the mock—and there may have been multiple calls in this case—matches my expectations:

```
from unittest.mock import ANY
```

```
mock = Mock(spec=get_animals)
mock('database 1', 'Rabbit')
mock('database 2', 'Bison')
mock('database 3', 'Meerkat')

mock.assert_called_with(ANY, 'Meerkat')
```

ANY is useful in tests when a parameter is not core to the behavior that's being tested. It's often worth erring on the side of under-specifying tests by using ANY more liberally instead of over-specifying tests and having to plumb through various test parameter expectations.

The Mock class also makes it easy to mock exceptions being raised:

```
class MyError(Exception):
    pass

mock = Mock(spec=get_animals)
mock.side_effect = MyError('Whoops! Big problem')
result = mock(database, 'Meerkat')

>>>
Traceback ...
MyError: Whoops! Big problem
```

There are many more features available, so be sure to see help(unittest.mock.Mock) for the full range of options.

Now that I've shown the mechanics of how a Mock works, I can apply it to an actual testing situation to show how to use it effectively in writing unit tests. Here, I define a function to do the rounds of feeding animals at the zoo, given a set of database-interacting functions:

```
def get_food_period(database, species):
    # Query the database
    ...
    # Return a time delta

def feed_animal(database, name, when):
    # Write to the database
    ...

def do_rounds(database, species):
    now = datetime.datetime.utcnow()
    feeding_timedelta = get_food_period(database, species)
    animals = get_animals(database, species)
    fed = 0
```

```
for name, last_mealtime in animals:
    if (now - last_mealtime) > feeding_timedelta:
        feed_animal(database, name, now)
        fed += 1

return fed
```

The goal of my test is to verify that when do_rounds is run, the right animals got fed, the latest feeding time was recorded to the database, and the total number of animals fed returned by the function matches the correct total. In order to do all this, I need to mock out datetime.utcnow so my tests have a stable time that isn't affected by daylight saving time and other ephemeral changes. I need to mock out get_food_period and get_animals to return values that would have come from the database. And I need to mock out feed_animal to accept data that would have been written back to the database.

The question is: Even if I know how to create these mock functions and set expectations, how do I get the do_rounds function that's being tested to use the mock dependent functions instead of the real versions? One approach is to inject everything as keyword-only arguments (see Item 25: "Enforce Clarity with Keyword-Only and Positional-Only Arguments"):

```
def do_rounds(database, species, *,
              now_func=datetime.utcnow,
              food_func=get_food_period,
              animals_func=get_animals,
              feed_func=feed_animal):
    now = now_func()
    feeding_timedelta = food_func(database, species)
    animals = animals_func(database, species)
    fed = 0

    for name, last_mealtime in animals:
        if (now - last_mealtime) > feeding_timedelta:
            feed_func(database, name, now)
            fed += 1

    return fed
```

To test this function, I need to create all of the Mock instances upfront and set their expectations:

```
from datetime import timedelta
```

```
now_func = Mock(spec=datetime.utcnow)
now_func.return_value = datetime(2019, 6, 5, 15, 45)

food_func = Mock(spec=get_food_period)
food_func.return_value = timedelta(hours=3)

animals_func = Mock(spec=get_animals)
animals_func.return_value = [
    ('Spot', datetime(2019, 6, 5, 11, 15)),
    ('Fluffy', datetime(2019, 6, 5, 12, 30)),
    ('Jojo', datetime(2019, 6, 5, 12, 45)),
]

feed_func = Mock(spec=feed_animal)
```

Then, I can run the test by passing the mocks into the do_rounds function to override the defaults:

```
result = do_rounds(
    database,
    'Meerkat',
    now_func=now_func,
    food_func=food_func,
    animals_func=animals_func,
    feed_func=feed_func)

assert result == 2
```

Finally, I can verify that all of the calls to dependent functions matched my expectations:

```
from unittest.mock import call

food_func.assert_called_once_with(database, 'Meerkat')

animals_func.assert_called_once_with(database, 'Meerkat')

feed_func.assert_has_calls(
    [
        call(database, 'Spot', now_func.return_value),
        call(database, 'Fluffy', now_func.return_value),
    ],
    any_order=True)
```

I don't verify the parameters to the datetime.utcnow mock or how many times it was called because that's indirectly verified by the return value of the function. For get_food_period and get_animals, I verify a single call with the specified parameters by using assert_called_once_with.

For the feed_animal function, I verify that two calls were made—and their order didn't matter—to write to the database using the unittest.mock.call helper and the assert_has_calls method.

This approach of using keyword-only arguments for injecting mocks works, but it's quite verbose and requires changing every function you want to test. The unittest.mock.patch family of functions makes injecting mocks easier. It temporarily reassigns an attribute of a module or class, such as the database-accessing functions that I defined above. For example, here I override get_animals to be a mock using patch:

```
from unittest.mock import patch

print('Outside patch:', get_animals)

with patch('__main__.get_animals'):
    print('Inside patch: ', get_animals)

print('Outside again:', get_animals)
>>>
Outside patch: <function get_animals at 0x109217040>
Inside patch:  <MagicMock name='get_animals' id='4454622832'>
Outside again: <function get_animals at 0x109217040>
```

patch works for many modules, classes, and attributes. It can be used in with statements (see Item 66: "Consider contextlib and with Statements for Reusable try/finally Behavior"), as a function decorator (see Item 26: "Define Function Decorators with functools.wraps"), or in the setUp and tearDown methods of TestCase classes (see Item 76: "Verify Related Behaviors in TestCase Subclasses"). For the full range of options, see help(unittest.mock.patch).

However, patch doesn't work in all cases. For example, to test do_rounds I need to mock out the current time returned by the datetime.utcnow class method. Python won't let me do that because the datetime class is defined in a C-extension module, which can't be modified in this way:

```
fake_now = datetime(2019, 6, 5, 15, 45)

with patch('datetime.datetime.utcnow'):
    datetime.utcnow.return_value = fake_now
>>>
Traceback ...
TypeError: can't set attributes of built-in/extension type
➥'datetime.datetime'
```

To work around this, I can create another helper function to fetch time that can be patched:

```
def get_do_rounds_time():
    return datetime.datetime.utcnow()


def do_rounds(database, species):
    now = get_do_rounds_time()
    ...


with patch('__main__.get_do_rounds_time'):
    ...
```

Alternatively, I can use a keyword-only argument for the datetime.utcnow mock and use patch for all of the other mocks:

```
def do_rounds(database, species, *, utcnow=datetime.utcnow):
    now = utcnow()
    feeding_timedelta = get_food_period(database, species)
    animals = get_animals(database, species)
    fed = 0

    for name, last_mealtime in animals:
        if (now - last_mealtime) > feeding_timedelta:
            feed_func(database, name, now)
            fed += 1

    return fed
```

I'm going to go with the latter approach. Now, I can use the patch.multiple function to create many mocks and set their expectations:

```
from unittest.mock import DEFAULT

with patch.multiple('__main__',
                    autospec=True,
                    get_food_period=DEFAULT,
                    get_animals=DEFAULT,
                    feed_animal=DEFAULT):
    now_func = Mock(spec=datetime.utcnow)
    now_func.return_value = datetime(2019, 6, 5, 15, 45)
    get_food_period.return_value = timedelta(hours=3)
    get_animals.return_value = [
        ('Spot', datetime(2019, 6, 5, 11, 15)),
        ('Fluffy', datetime(2019, 6, 5, 12, 30)),
        ('Jojo', datetime(2019, 6, 5, 12, 45))
    ]
```

With the setup ready, I can run the test and verify that the calls were correct inside the with statement that used patch.multiple:

```
result = do_rounds(database, 'Meerkat', utcnow=now_func)
assert result == 2

food_func.assert_called_once_with(database, 'Meerkat')
animals_func.assert_called_once_with(database, 'Meerkat')
feed_func.assert_has_calls(
    [
        call(database, 'Spot', now_func.return_value),
        call(database, 'Fluffy', now_func.return_value),
    ],
    any_order=True)
```

The keyword arguments to patch.multiple correspond to names in the __main__ module that I want to override during the test. The DEFAULT value indicates that I want a standard Mock instance to be created for each name. All of the generated mocks will adhere to the specification of the objects they are meant to simulate, thanks to the autospec=True parameter.

These mocks work as expected, but it's important to realize that it's possible to further improve the readability of these tests and reduce boilerplate by refactoring your code to be more testable (see Item 79: "Encapsulate Dependencies to Facilitate Mocking and Testing").

Things to Remember

✦ The unittest.mock module provides a way to simulate the behavior of interfaces using the Mock class. Mocks are useful in tests when it's difficult to set up the dependencies that are required by the code that's being tested.

✦ When using mocks, it's important to verify both the behavior of the code being tested and how dependent functions were called by that code, using the Mock.assert_called_once_with family of methods.

✦ Keyword-only arguments and the unittest.mock.patch family of functions can be used to inject mocks into the code being tested.

Item 79: Encapsulate Dependencies to Facilitate Mocking and Testing

In the previous item (see Item 78: "Use Mocks to Test Code with Complex Dependencies"), I showed how to use the facilities of the unittest.mock built-in module—including the Mock class and patch family of functions—to write tests that have complex dependencies, such as a database. However, the resulting test code requires a lot of boilerplate, which could make it more difficult for new readers of the code to understand what the tests are trying to verify.

One way to improve these tests is to use a wrapper object to encapsulate the database's interface instead of passing a DatabaseConnection object to functions as an argument. It's often worth refactoring your code (see Item 89: "Consider warnings to Refactor and Migrate Usage" for one approach) to use better abstractions because it facilitates creating mocks and writing tests. Here, I redefine the various database helper functions from the previous item as methods on a class instead of as independent functions:

```
class ZooDatabase:
    ...

    def get_animals(self, species):
        ...

    def get_food_period(self, species):
        ...

    def feed_animal(self, name, when):
        ...
```

Now, I can redefine the do_rounds function to call methods on a ZooDatabase object:

```
from datetime import datetime

def do_rounds(database, species, *, utcnow=datetime.utcnow):
    now = utcnow()
    feeding_timedelta = database.get_food_period(species)
    animals = database.get_animals(species)
    fed = 0

    for name, last_mealtime in animals:
        if (now - last_mealtime) >= feeding_timedelta:
            database.feed_animal(name, now)
            fed += 1

    return fed
```

Writing a test for do_rounds is now a lot easier because I no longer need to use unittest.mock.patch to inject the mock into the code being tested. Instead, I can create a Mock instance to represent a ZooDatabase and pass that in as the database parameter. The Mock class returns a mock object for any attribute name that is accessed. Those attributes can be called like methods, which I can then use to set expectations and verify calls. This makes it easy to mock out all of the methods of a class:

```
from unittest.mock import Mock

database = Mock(spec=ZooDatabase)
print(database.feed_animal)
database.feed_animal()
database.feed_animal.assert_any_call()
```

```
>>>
<Mock name='mock.feed_animal' id='4384773408'>
```

I can rewrite the Mock setup code by using the ZooDatabase encapsulation:

```
from datetime import timedelta
from unittest.mock import call

now_func = Mock(spec=datetime.utcnow)
now_func.return_value = datetime(2019, 6, 5, 15, 45)

database = Mock(spec=ZooDatabase)
database.get_food_period.return_value = timedelta(hours=3)
database.get_animals.return_value = [
    ('Spot', datetime(2019, 6, 5, 11, 15)),
    ('Fluffy', datetime(2019, 6, 5, 12, 30)),
    ('Jojo', datetime(2019, 6, 5, 12, 55))
]
```

Then I can run the function being tested and verify that all dependent methods were called as expected:

```
result = do_rounds(database, 'Meerkat', utcnow=now_func)
assert result == 2

database.get_food_period.assert_called_once_with('Meerkat')
database.get_animals.assert_called_once_with('Meerkat')
database.feed_animal.assert_has_calls(
    [
        call('Spot', now_func.return_value),
        call('Fluffy', now_func.return_value),
    ],
    any_order=True)
```

Using the spec parameter to Mock is especially useful when mocking classes because it ensures that the code under test doesn't call a misspelled method name by accident. This allows you to avoid a common pitfall where the same bug is present in both the code and the unit test, masking a real error that will later reveal itself in production:

```
database.bad_method_name()
```

```
>>>
Traceback ...
AttributeError: Mock object has no attribute 'bad_method_name'
```

If I want to test this program end-to-end with a mid-level integration test (see Item 77: "Isolate Tests from Each Other with setUp, tearDown, setUpModule, and tearDownModule"), I still need a way to inject a mock ZooDatabase into the program. I can do this by creating a helper function that acts as a seam for *dependency injection*. Here, I define such a helper function that caches a ZooDatabase in module scope (see Item 86: "Consider Module-Scoped Code to Configure Deployment Environments") by using a global statement:

```
DATABASE = None

def get_database():
    global DATABASE
    if DATABASE is None:
        DATABASE = ZooDatabase()
    return DATABASE

def main(argv):
    database = get_database()
    species = argv[1]
    count = do_rounds(database, species)
    print(f'Fed {count} {species}(s)')
    return 0
```

Now, I can inject the mock ZooDatabase using patch, run the test, and verify the program's output. I'm not using a mock datetime.utcnow here; instead, I'm relying on the database records returned by the mock to be relative to the current time in order to produce similar behavior to the unit test. This approach is more flaky than mocking everything, but it also tests more surface area:

```
import contextlib
import io
from unittest.mock import patch
```

```
with patch('__main__.DATABASE', spec=ZooDatabase):
    now = datetime.utcnow()

    DATABASE.get_food_period.return_value = timedelta(hours=3)
    DATABASE.get_animals.return_value = [
        ('Spot', now - timedelta(minutes=4.5)),
        ('Fluffy', now - timedelta(hours=3.25)),
        ('Jojo', now - timedelta(hours=3)),
    ]

    fake_stdout = io.StringIO()
    with contextlib.redirect_stdout(fake_stdout):
        main(['program name', 'Meerkat'])

    found = fake_stdout.getvalue()
    expected = 'Fed 2 Meerkat(s)\n'

    assert found == expected
```

The results match my expectations. Creating this integration test was straightforward because I designed the implementation to make it easier to test.

Things to Remember

✦ When unit tests require a lot of repeated boilerplate to set up mocks, one solution may be to encapsulate the functionality of dependencies into classes that are more easily mocked.

✦ The Mock class of the unittest.mock built-in module simulates classes by returning a new mock, which can act as a mock method, for each attribute that is accessed.

✦ For end-to-end tests, it's valuable to refactor your code to have more helper functions that can act as explicit seams to use for injecting mock dependencies in tests.

Item 80: Consider Interactive Debugging with pdb

Everyone encounters bugs in code while developing programs. Using the print function can help you track down the sources of many issues (see Item 75: "Use repr Strings for Debugging Output"). Writing tests for specific cases that cause trouble is another great way to isolate problems (see Item 76: "Verify Related Behaviors in TestCase Subclasses").

But these tools aren't enough to find every root cause. When you need something more powerful, it's time to try Python's built-in *interactive debugger*. The debugger lets you inspect program state, print local variables, and step through a Python program one statement at a time.

In most other programming languages, you use a debugger by specifying what line of a source file you'd like to stop on, and then execute the program. In contrast, with Python, the easiest way to use the debugger is by modifying your program to directly initiate the debugger just before you think you'll have an issue worth investigating. This means that there is no difference between starting a Python program in order to run the debugger and starting it normally.

To initiate the debugger, all you have to do is call the breakpoint built-in function. This is equivalent to importing the pdb built-in module and running its set_trace function:

```python
# always_breakpoint.py
import math

def compute_rmse(observed, ideal):
    total_err_2 = 0
    count = 0
    for got, wanted in zip(observed, ideal):
        err_2 = (got - wanted) ** 2
        breakpoint()  # Start the debugger here
        total_err_2 += err_2
        count += 1

    mean_err = total_err_2 / count
    rmse = math.sqrt(mean_err)
    return rmse

result = compute_rmse(
    [1.8, 1.7, 3.2, 6],
    [2, 1.5, 3, 5])
print(result)
```

As soon as the breakpoint function runs, the program pauses its execution before the line of code immediately following the breakpoint call. The terminal that started the program turns into an interactive Python shell:

```
$ python3 always_breakpoint.py
> always_breakpoint.py(12)compute_rmse()
-> total_err_2 += err_2
(Pdb)
```

At the (Pdb) prompt, you can type in the names of local variables to see their values printed out (or use p <name>). You can see a list of all local variables by calling the locals built-in function. You can import modules, inspect global state, construct new objects, run the help built-in function, and even modify parts of the running program—whatever you need to do to aid in your debugging.

In addition, the debugger has a variety of special commands to control and understand program execution; type help to see the full list.

Three very useful commands make inspecting the running program easier:

- where: Print the current execution call stack. This lets you figure out where you are in your program and how you arrived at the breakpoint trigger.

- up: Move your scope up the execution call stack to the caller of the current function. This allows you to inspect the local variables in higher levels of the program that led to the breakpoint.

- down: Move your scope back down the execution call stack one level.

When you're done inspecting the current state, you can use these five debugger commands to control the program's execution in different ways:

- step: Run the program until the next line of execution in the program, and then return control back to the debugger prompt. If the next line of execution includes calling a function, the debugger stops within the function that was called.

- next: Run the program until the next line of execution in the current function, and then return control back to the debugger prompt. If the next line of execution includes calling a function, the debugger will not stop until the called function has returned.

- return: Run the program until the current function returns, and then return control back to the debugger prompt.

- continue: Continue running the program until the next breakpoint is hit (either through the breakpoint call or one added by a debugger command).

- quit: Exit the debugger and end the program. Run this command if you've found the problem, gone too far, or need to make program modifications and try again.

The breakpoint function can be called anywhere in a program. If you know that the problem you're trying to debug happens only under special circumstances, then you can just write plain old Python code to call breakpoint after a specific condition is met. For example, here I start the debugger only if the squared error for a datapoint is more than 1:

```
# conditional_breakpoint.py
def compute_rmse(observed, ideal):
    ...
    for got, wanted in zip(observed, ideal):
        err_2 = (got - wanted) ** 2
        if err_2 >= 1:  # Start the debugger if True
            breakpoint()
        total_err_2 += err_2
        count += 1
    ...
result = compute_rmse(
    [1.8, 1.7, 3.2, 7],
    [2, 1.5, 3, 5])
print(result)
```

When I run the program and it enters the debugger, I can confirm that the condition was true by inspecting local variables:

```
$ python3 conditional_breakpoint.py
> conditional_breakpoint.py(14)compute_rmse()
-> total_err_2 += err_2
(Pdb) wanted
5
(Pdb) got
7
(Pdb) err_2
4
```

Another useful way to reach the debugger prompt is by using *post-mortem debugging*. This enables you to debug a program *after* it's already raised an exception and crashed. This is especially helpful when you're not quite sure where to put the breakpoint function call.

Here, I have a script that will crash due to the 7j complex number being present in one of the function's arguments:

```
# postmortem_breakpoint.py
import math

def compute_rmse(observed, ideal):
    ...
```

```
result = compute_rmse(
    [1.8, 1.7, 3.2, 7j],  # Bad input
    [2, 1.5, 3, 5])
print(result)
```

I use the command line python3 -m pdb -c continue <program path> to run the program under control of the pdb module. The continue command tells pdb to get the program started immediately. Once it's running, the program hits a problem and automatically enters the interactive debugger, at which point I can inspect the program state:

```
$ python3 -m pdb -c continue postmortem_breakpoint.py
Traceback (most recent call last):
  File ".../pdb.py", line 1697, in main
    pdb._runscript(mainpyfile)
  File ".../pdb.py", line 1566, in _runscript
    self.run(statement)
  File ".../bdb.py", line 585, in run
    exec(cmd, globals, locals)
  File "<string>", line 1, in <module>
  File "postmortem_breakpoint.py", line 4, in <module>
    import math
  File "postmortem_breakpoint.py", line 16, in compute_rmse
    rmse = math.sqrt(mean_err)
TypeError: can't convert complex to float
Uncaught exception. Entering post mortem debugging
Running 'cont' or 'step' will restart the program
> postmortem_breakpoint.py(16)compute_rmse()
-> rmse = math.sqrt(mean_err)
(Pdb) mean_err
(-5.97-17.5j)
```

You can also use post-mortem debugging after hitting an uncaught exception in the interactive Python interpreter by calling the pm function of the pdb module (which is often done in a single line as import pdb; pdb.pm()):

```
$ python3
>>> import my_module
>>> my_module.compute_stddev([5])
Traceback (most recent call last):
  File "<stdin>", line 1, in <module>
  File "my_module.py", line 17, in compute_stddev
    variance = compute_variance(data)
  File "my_module.py", line 13, in compute_variance
    variance = err_2_sum / (len(data) - 1)
```

```
ZeroDivisionError: float division by zero
>>> import pdb; pdb.pm()
> my_module.py(13)compute_variance()
-> variance = err_2_sum / (len(data) - 1)
(Pdb) err_2_sum
0.0
(Pdb) len(data)
1
```

Things to Remember

✦ You can initiate the Python interactive debugger at a point of interest directly in your program by calling the breakpoint built-in function.

✦ The Python debugger prompt is a full Python shell that lets you inspect and modify the state of a running program.

✦ pdb shell commands let you precisely control program execution and allow you to alternate between inspecting program state and progressing program execution.

✦ The pdb module can be used for debug exceptions after they happen in independent Python programs (using python -m pdb -c continue <program path>) or the interactive Python interpreter (using import pdb; pdb.pm()).

Item 81: Use tracemalloc to Understand Memory Usage and Leaks

Memory management in the default implementation of Python, CPython, uses reference counting. This ensures that as soon as all references to an object have expired, the referenced object is also cleared from memory, freeing up that space for other data. CPython also has a built-in cycle detector to ensure that self-referencing objects are eventually garbage collected.

In theory, this means that most Python programmers don't have to worry about allocating or deallocating memory in their programs. It's taken care of automatically by the language and the CPython runtime. However, in practice, programs eventually do run out of memory due to no longer useful references still being held. Figuring out where a Python program is using or leaking memory proves to be a challenge.

The first way to debug memory usage is to ask the gc built-in module to list every object currently known by the garbage collector. Although it's quite a blunt tool, this approach lets you quickly get a sense of where your program's memory is being used.

Here, I define a module that fills up memory by keeping references:

```
# waste_memory.py
import os

class MyObject:
    def __init__(self):
        self.data = os.urandom(100)

def get_data():
    values = []
    for _ in range(100):
        obj = MyObject()
        values.append(obj)
    return values

def run():
    deep_values = []
    for _ in range(100):
        deep_values.append(get_data())
    return deep_values
```

Then, I run a program that uses the gc built-in module to print out how many objects were created during execution, along with a small sample of allocated objects:

```
# using_gc.py
import gc

found_objects = gc.get_objects()
print('Before:', len(found_objects))

import waste_memory

hold_reference = waste_memory.run()

found_objects = gc.get_objects()
print('After: ', len(found_objects))
for obj in found_objects[:3]:
    print(repr(obj)[:100])

>>>
Before: 6207
After:  16801
```

```
<waste_memory.MyObject object at 0x10390aeb8>
<waste_memory.MyObject object at 0x10390aef0>
<waste_memory.MyObject object at 0x10390af28>
...
```

The problem with `gc.get_objects` is that it doesn't tell you anything about *how* the objects were allocated. In complicated programs, objects of a specific class could be allocated many different ways. Knowing the overall number of objects isn't nearly as important as identifying the code responsible for allocating the objects that are leaking memory.

Python 3.4 introduced a new `tracemalloc` built-in module for solving this problem. `tracemalloc` makes it possible to connect an object back to where it was allocated. You use it by taking before and after snapshots of memory usage and comparing them to see what's changed. Here, I use this approach to print out the top three memory usage offenders in a program:

```python
# top_n.py
import tracemalloc

tracemalloc.start(10)                        # Set stack depth
time1 = tracemalloc.take_snapshot()          # Before snapshot

import waste_memory

x = waste_memory.run()                       # Usage to debug
time2 = tracemalloc.take_snapshot()          # After snapshot

stats = time2.compare_to(time1, 'lineno')    # Compare snapshots
for stat in stats[:3]:
    print(stat)

>>>
waste_memory.py:5: size=2392 KiB (+2392 KiB), count=29994
➥(+29994), average=82 B
waste_memory.py:10: size=547 KiB (+547 KiB), count=10001
➥(+10001), average=56 B
waste_memory.py:11: size=82.8 KiB (+82.8 KiB), count=100
➥(+100), average=848 B
```

The size and count labels in the output make it immediately clear which objects are dominating my program's memory usage and where in the source code they were allocated.

The tracemalloc module can also print out the full stack trace of each allocation (up to the number of frames passed to the tracemalloc.start function). Here, I print out the stack trace of the biggest source of memory usage in the program:

```
# with_trace.py
import tracemalloc

tracemalloc.start(10)
time1 = tracemalloc.take_snapshot()

import waste_memory

x = waste_memory.run()
time2 = tracemalloc.take_snapshot()

stats = time2.compare_to(time1, 'traceback')
top = stats[0]
print('Biggest offender is:')
print('\n'.join(top.traceback.format()))

>>>
Biggest offender is:
  File "with_trace.py", line 9
    x = waste_memory.run()
  File "waste_memory.py", line 17
    deep_values.append(get_data())
  File "waste_memory.py", line 10
    obj = MyObject()
  File "waste_memory.py", line 5
    self.data = os.urandom(100)
```

A stack trace like this is most valuable for figuring out which particular usage of a common function or class is responsible for memory consumption in a program.

Things to Remember

✦ It can be difficult to understand how Python programs use and leak memory.

✦ The gc module can help you understand which objects exist, but it has no information about how they were allocated.

✦ The tracemalloc built-in module provides powerful tools for understanding the sources of memory usage.

Chapter 10   Collaboration

Python has language features that help you construct well-defined APIs with clear interface boundaries. The Python community has established best practices to maximize the maintainability of code over time. In addition, some standard tools that ship with Python enable large teams to work together across disparate environments.

Collaborating with others on Python programs requires being deliberate in how you write your code. Even if you're working on your own, chances are you'll be using code written by someone else via the standard library or open source packages. It's important to understand the mechanisms that make it easy to collaborate with other Python programmers.

Item 82: Know Where to Find Community-Built Modules

Python has a central repository of modules that you can install and use in your programs. These modules are built and maintained by people like you: the Python community. When you find yourself facing an unfamiliar challenge, the Python Package Index (PyPI) is a great place to look for code that will get you closer to your goal.

To use the Package Index, you need to use the command-line tool pip (a recursive acronym for "pip installs packages"). pip can be run with python3 -m pip to ensure that packages are installed for the correct version of Python on your system (see Item 1: "Know Which Version of Python You're Using"). Using pip to install a new module is simple. For example, here I install the pytz module that I use elsewhere in this book (see Item 67: "Use datetime Instead of time for Local Clocks"):

```
$ python3 -m pip install pytz
Collecting pytz
```

```
  Downloading ...
Installing collected packages: pytz
Successfully installed pytz-2018.9
```

pip is best used together with the built-in module venv to consistently track sets of packages to install for your projects (see Item 83: "Use Virtual Environments for Isolated and Reproducible Dependencies"). You can also create your own PyPI packages to share with the Python community or host your own private package repositories for use with pip.

Each module in the PyPI has its own software license. Most of the packages, especially the popular ones, have free or open source licenses. In most cases, these licenses allow you to include a copy of the module with your program; when in doubt, talk to a lawyer.

Things to Remember

✦ The Python Package Index (PyPI) contains a wealth of common packages that are built and maintained by the Python community.

✦ pip is the command-line tool you can use to install packages from PyPI.

✦ The majority of PyPI modules are free and open source software.

Item 83: Use Virtual Environments for Isolated and Reproducible Dependencies

Building larger and more complex programs often leads you to rely on various packages from the Python community (see Item 82: "Know Where to Find Community-Built Modules"). You'll find yourself running the python3 -m pip command-line tool to install packages like pytz, numpy, and many others.

The problem is that, by default, pip installs new packages in a global location. That causes all Python programs on your system to be affected by these installed modules. In theory, this shouldn't be an issue. If you install a package and never import it, how could it affect your programs?

The trouble comes from transitive dependencies: the packages that the packages you install depend on. For example, you can see what the Sphinx package depends on after installing it by asking pip:

```
$ python3 -m pip show Sphinx
Name: Sphinx
```

```
Version: 2.1.2
Summary: Python documentation generator
Location: /usr/local/lib/python3.8/site-packages
Requires: alabaster, imagesize, requests,
➥sphinxcontrib-applehelp, sphinxcontrib-qthelp,
➥Jinja2, setuptools, sphinxcontrib-jsmath,
➥sphinxcontrib-serializinghtml, Pygments, snowballstemmer,
➥packaging, sphinxcontrib-devhelp, sphinxcontrib-htmlhelp,
➥babel, docutils
Required-by:
```

If you install another package like flask, you can see that it, too, depends on the Jinja2 package:

```
$ python3 -m pip show flask
Name: Flask
Version: 1.0.3
Summary: A simple framework for building complex web applications.
Location: /usr/local/lib/python3.8/site-packages
Requires: itsdangerous, click, Jinja2, Werkzeug
Required-by:
```

A dependency conflict can arise as Sphinx and flask diverge over time. Perhaps right now they both require the same version of Jinja2, and everything is fine. But six months or a year from now, Jinja2 may release a new version that makes breaking changes to users of the library. If you update your global version of Jinja2 with python3 -m pip install --upgrade Jinja2, you may find that Sphinx breaks, while flask keeps working.

The cause of such breakage is that Python can have only a single global version of a module installed at a time. If one of your installed packages must use the new version and another package must use the old version, your system isn't going to work properly; this situation is often called *dependency hell*.

Such breakage can even happen when package maintainers try their best to preserve API compatibility between releases (see Item 85: "Use Packages to Organize Modules and Provide Stable APIs"). New versions of a library can subtly change behaviors that API-consuming code relies on. Users on a system may upgrade one package to a new version but not others, which could break dependencies. If you're not careful there's a constant risk of the ground moving beneath your feet.

These difficulties are magnified when you collaborate with other developers who do their work on separate computers. It's best to assume the worst: that the versions of Python and global packages that they have installed on their machines will be slightly different from yours. This can cause frustrating situations such as a codebase working perfectly on one programmer's machine and being completely broken on another's.

The solution to all of these problems is using a tool called venv, which provides *virtual environments*. Since Python 3.4, pip and the venv module have been available by default along with the Python installation (accessible with python -m venv).

venv allows you to create isolated versions of the Python environment. Using venv, you can have many different versions of the same package installed on the same system at the same time without conflicts. This means you can work on many different projects and use many different tools on the same computer. venv does this by installing explicit versions of packages and their dependencies into completely separate directory structures. This makes it possible to reproduce a Python environment that you know will work with your code. It's a reliable way to avoid surprising breakages.

Using venv on the Command Line

Here's a quick tutorial on how to use venv effectively. Before using the tool, it's important to note the meaning of the python3 command line on your system. On my computer, python3 is located in the /usr/local/bin directory and evaluates to version 3.8.0 (see Item 1: "Know Which Version of Python You're Using"):

```
$ which python3
/usr/local/bin/python3
$ python3 --version
Python 3.8.0
```

To demonstrate the setup of my environment, I can test that running a command to import the pytz module doesn't cause an error. This works because I already have the pytz package installed as a global module:

```
$ python3 -c 'import pytz'
$
```

Now, I use venv to create a new virtual environment called myproject. Each virtual environment must live in its own unique directory. The result of the command is a tree of directories and files that are used to manage the virtual environment:

```
$ python3 -m venv myproject
$ cd myproject
```

```
$ ls
bin      include      lib      pyvenv.cfg
```

To start using the virtual environment, I use the source command from my shell on the bin/activate script. activate modifies all of my environment variables to match the virtual environment. It also updates my command-line prompt to include the virtual environment name ("myproject") to make it extremely clear what I'm working on:

```
$ source bin/activate
(myproject)$
```

On Windows the same script is available as:

```
C:\> myproject\Scripts\activate.bat
(myproject) C:>
```

Or with PowerShell as:

```
PS C:\> myproject\Scripts\activate.ps1
(myproject) PS C:>
```

After activation, the path to the python3 command-line tool has moved to within the virtual environment directory:

```
(myproject)$ which python3
/tmp/myproject/bin/python3
(myproject)$ ls -l /tmp/myproject/bin/python3
... -> /usr/local/bin/python3.8
```

This ensures that changes to the outside system will not affect the virtual environment. Even if the outer system upgrades its default python3 to version 3.9, my virtual environment will still explicitly point to version 3.8.

The virtual environment I created with venv starts with no packages installed except for pip and setuptools. Trying to use the pytz package that was installed as a global module in the outside system will fail because it's unknown to the virtual environment:

```
(myproject)$ python3 -c 'import pytz'
Traceback (most recent call last):
  File "<string>", line 1, in <module>
ModuleNotFoundError: No module named 'pytz'
```

I can use the pip command-line tool to install the pytz module into my virtual environment:

```
(myproject)$ python3 -m pip install pytz
Collecting pytz
```

```
  Downloading ...
Installing collected packages: pytz
Successfully installed pytz-2019.1
```

Once it's installed, I can verify that it's working by using the same test import command:

```
(myproject)$ python3 -c 'import pytz'
(myproject)$
```

When I'm done with a virtual environment and want to go back to my default system, I use the deactivate command. This restores my environment to the system defaults, including the location of the python3 command-line tool:

```
(myproject)$ which python3
/tmp/myproject/bin/python3
(myproject)$ deactivate
$ which python3
/usr/local/bin/python3
```

If I ever want to work in the myproject environment again, I can just run source bin/activate in the directory as before.

Reproducing Dependencies

Once you are in a virtual environment, you can continue installing packages in it with pip as you need them. Eventually, you might want to copy your environment somewhere else. For example, say that I want to reproduce the development environment from my workstation on a server in a datacenter. Or maybe I want to clone someone else's environment on my own machine so I can help debug their code.

venv makes such tasks easy. I can use the python3 -m pip freeze command to save all of my explicit package dependencies into a file (which, by convention, is named requirements.txt):

```
(myproject)$ python3 -m pip freeze > requirements.txt
(myproject)$ cat requirements.txt
certifi==2019.3.9
chardet==3.0.4
idna==2.8
numpy==1.16.2
pytz==2018.9
requests==2.21.0
urllib3==1.24.1
```

Now, imagine that I'd like to have another virtual environment that matches the myproject environment. I can create a new directory as before by using venv and activate it:

```
$ python3 -m venv otherproject
$ cd otherproject
$ source bin/activate
(otherproject)$
```

The new environment will have no extra packages installed:

```
(otherproject)$ python3 -m pip list
Package    Version
---------- -------
pip        10.0.1
setuptools 39.0.1
```

I can install all of the packages from the first environment by running python3 -m pip install on the requirements.txt that I generated with the python3 -m pip freeze command:

```
(otherproject)$ python3 -m pip install -r /tmp/myproject/
➥requirements.txt
```

This command cranks along for a little while as it retrieves and installs all of the packages required to reproduce the first environment. When it's done, I can list the set of installed packages in the second virtual environment and should see the same list of dependencies found in the first virtual environment:

```
(otherproject)$ python3 -m pip list
Package    Version
---------- --------
certifi    2019.3.9
chardet    3.0.4
idna       2.8
numpy      1.16.2
pip        10.0.1
pytz       2018.9
requests   2.21.0
setuptools 39.0.1
urllib3    1.24.1
```

Using a requirements.txt file is ideal for collaborating with others through a revision control system. You can commit changes to your code at the same time you update your list of package dependencies, ensuring that they move in lockstep. However, it's important to note that the specific version of Python you're using is *not* included in the requirements.txt file, so that must be managed separately.

The gotcha with virtual environments is that moving them breaks everything because all of the paths, like the python3 command-line tool, are hard-coded to the environment's install directory. But ultimately this limitation doesn't matter. The whole purpose of virtual environments is to make it easy to reproduce a setup. Instead of moving a virtual environment directory, just use python3 -m pip freeze on the old one, create a new virtual environment somewhere else, and reinstall everything from the requirements.txt file.

Things to Remember

✦ Virtual environments allow you to use pip to install many different versions of the same package on the same machine without conflicts.

✦ Virtual environments are created with python -m venv, enabled with source bin/activate, and disabled with deactivate.

✦ You can dump all of the requirements of an environment with python3 -m pip freeze. You can reproduce an environment by running python3 -m pip install -r requirements.txt.

Item 84: Write Docstrings for Every Function, Class, and Module

Documentation in Python is extremely important because of the dynamic nature of the language. Python provides built-in support for attaching documentation to blocks of code. Unlike with many other languages, the documentation from a program's source code is directly accessible as the program runs.

For example, you can add documentation by providing a *docstring* immediately after the def statement of a function:

```
def palindrome(word):
    """Return True if the given word is a palindrome."""
    return word == word[::-1]

assert palindrome('tacocat')
assert not palindrome('banana')
```

You can retrieve the docstring from within the Python program by accessing the function's __doc__ special attribute:

```
print(repr(palindrome.__doc__))

>>>
'Return True if the given word is a palindrome.'
```

You can also use the built-in pydoc module from the command line to run a local web server that hosts all of the Python documentation that's accessible to your interpreter, including modules that you've written:

```
$ python3 -m pydoc -p 1234
Server ready at http://localhost:1234/
Server commands: [b]rowser, [q]uit
server> b
```

Docstrings can be attached to functions, classes, and modules. This connection is part of the process of compiling and running a Python program. Support for docstrings and the __doc__ attribute has three consequences:

- The accessibility of documentation makes interactive develop-ment easier. You can inspect functions, classes, and modules to see their documentation by using the help built-in function. This makes the Python interactive interpreter (the Python "shell") and tools like IPython Notebook a joy to use while you're developing algorithms, testing APIs, and writing code snippets.

- A standard way of defining documentation makes it easy to build tools that convert the text into more appealing formats (like HTML). This has led to excellent documentation-generation tools for the Python community, such as Sphinx. It has also enabled community-funded sites like Read the Docs that provide free hosting of beautiful-looking documentation for open source Python projects.

- Python's first-class, accessible, and good-looking documentation encourages people to write more documentation. The members of the Python community have a strong belief in the importance of documentation. There's an assumption that "good code" also means well-documented code. This means that you can expect most open source Python libraries to have decent documentation.

To participate in this excellent culture of documentation, you need to follow a few guidelines when you write docstrings. The full details are discussed online in PEP 257. There are a few best practices you should be sure to follow.

Documenting Modules

Each module should have a top-level docstring—a string literal that is the first statement in a source file. It should use three double quotes ("""). The goal of this docstring is to introduce the module and its contents.

The first line of the docstring should be a single sentence describing the module's purpose. The paragraphs that follow should contain the details that all users of the module should know about its operation. The module docstring is also a jumping-off point where you can highlight important classes and functions found in the module.

Here's an example of a module docstring:

```
# words.py
#!/usr/bin/env python3
"""Library for finding linguistic patterns in words.

Testing how words relate to each other can be tricky sometimes!
This module provides easy ways to determine when words you've
found have special properties.

Available functions:
- palindrome: Determine if a word is a palindrome.
- check_anagram: Determine if two words are anagrams.
...
"""

...
```

If the module is a command-line utility, the module docstring is also a great place to put usage information for running the tool.

Documenting Classes

Each class should have a class-level docstring. This largely follows the same pattern as the module-level docstring. The first line is the single-sentence purpose of the class. Paragraphs that follow discuss important details of the class's operation.

Important public attributes and methods of the class should be highlighted in the class-level docstring. It should also provide guidance to subclasses on how to properly interact with protected attributes (see Item 42: "Prefer Public Attributes Over Private Ones") and the superclass's methods.

Here's an example of a class docstring:

```
class Player:
    """Represents a player of the game.

    Subclasses may override the 'tick' method to provide
    custom animations for the player's movement depending
    on their power level, etc.
```

```
Public attributes:
- power: Unused power-ups (float between 0 and 1).
- coins: Coins found during the level (integer).
"""

...
```

Documenting Functions

Each public function and method should have a docstring. This follows the same pattern as the docstrings for modules and classes. The first line is a single-sentence description of what the function does. The paragraphs that follow should describe any specific behaviors and the arguments for the function. Any return values should be mentioned. Any exceptions that callers must handle as part of the function's interface should be explained (see Item 20: "Prefer Raising Exceptions to Returning None" for how to document raised exceptions).

Here's an example of a function docstring:

```
def find_anagrams(word, dictionary):
    """Find all anagrams for a word.

    This function only runs as fast as the test for
    membership in the 'dictionary' container.

    Args:
        word: String of the target word.
        dictionary: collections.abc.Container with all
            strings that are known to be actual words.

    Returns:
        List of anagrams that were found. Empty if
        none were found.
    """

    ...
```

There are also some special cases in writing docstrings for functions that are important to know:

- If a function has no arguments and a simple return value, a single-sentence description is probably good enough.

- If a function doesn't return anything, it's better to leave out any mention of the return value instead of saying "returns None."

- If a function's interface includes raising exceptions (see Item 20: "Prefer Raising Exceptions to Returning None" for an example), its docstring should describe each exception that's raised and when it's raised.

- If you don't expect a function to raise an exception during normal operation, don't mention that fact.

- If a function accepts a variable number of arguments (see Item 22: "Reduce Visual Noise with Variable Positional Arguments") or keyword arguments (see Item 23: "Provide Optional Behavior with Keyword Arguments"), use *args and **kwargs in the documented list of arguments to describe their purpose.

- If a function has arguments with default values, those defaults should be mentioned (see Item 24: "Use None and Docstrings to Specify Dynamic Default Arguments").

- If a function is a generator (see Item 30: "Consider Generators Instead of Returning Lists"), its docstring should describe what the generator yields when it's iterated.

- If a function is an asynchronous coroutine (see Item 60: "Achieve Highly Concurrent I/O with Coroutines"), its docstring should explain when it will stop execution.

Using Docstrings and Type Annotations

Python now supports type annotations for a variety of purposes (see Item 90: "Consider Static Analysis via typing to Obviate Bugs" for how to use them). The information they contain may be redundant with typical docstrings. For example, here is the function signature for find_anagrams with type annotations applied:

```
from typing import Container, List


def find_anagrams(word: str,
                  dictionary: Container[str]) -> List[str]:
    ...
```

There is no longer a need to specify in the docstring that the word argument is a string, since the type annotation has that information. The same goes for the dictionary argument being a collections.abc.Container. There's no reason to mention that the return type will be a list, since this fact is clearly annotated. And when no anagrams are found, the return value still must be a list, so it's implied that it will be empty; that doesn't need to be noted in the docstring. Here, I write the same function signature from above along with the docstring that has been shortened accordingly:

```
def find_anagrams(word: str,
                  dictionary: Container[str]) -> List[str]:
    """Find all anagrams for a word.
```

```
This function only runs as fast as the test for
membership in the 'dictionary' container.

Args:
    word: Target word.
    dictionary: All known actual words.

Returns:
    Anagrams that were found.
"""

...
```

The redundancy between type annotations and docstrings should be similarly avoided for instance fields, class attributes, and methods. It's best to have type information in only one place so there's less risk that it will skew from the actual implementation.

Things to Remember

✦ Write documentation for every module, class, method, and function using docstrings. Keep them up-to-date as your code changes.

✦ For modules: Introduce the contents of a module and any important classes or functions that all users should know about.

✦ For classes: Document behavior, important attributes, and subclass behavior in the docstring following the class statement.

✦ For functions and methods: Document every argument, returned value, raised exception, and other behaviors in the docstring following the def statement.

✦ If you're using type annotations, omit the information that's already present in type annotations from docstrings since it would be redundant to have it in both places.

Item 85: Use Packages to Organize Modules and Provide Stable APIs

As the size of a program's codebase grows, it's natural for you to reorganize its structure. You'll split larger functions into smaller functions. You'll refactor data structures into helper classes (see Item 37: "Compose Classes Instead of Nesting Many Levels of Built-in Types" for an example). You'll separate functionality into various modules that depend on each other.

At some point, you'll find yourself with so many modules that you need another layer in your program to make it understandable. For this purpose, Python provides *packages*. Packages are modules that contain other modules.

In most cases, packages are defined by putting an empty file named __init__.py into a directory. Once __init__.py is present, any other Python files in that directory will be available for import, using a path relative to the directory. For example, imagine that I have the following directory structure in my program:

```
main.py
mypackage/__init__.py
mypackage/models.py
mypackage/utils.py
```

To import the utils module, I use the absolute module name that includes the package directory's name:

```
# main.py
from mypackage import utils
```

This pattern continues when I have package directories present within other packages (like mypackage.foo.bar).

The functionality provided by packages has two primary purposes in Python programs.

Namespaces

The first use of packages is to help divide your modules into separate namespaces. They enable you to have many modules with the same filename but different absolute paths that are unique. For example, here's a program that imports attributes from two modules with the same filename, utils.py:

```
# main.py
from analysis.utils import log_base2_bucket
from frontend.utils import stringify

bucket = stringify(log_base2_bucket(33))
```

This approach breaks when the functions, classes, or submodules defined in packages have the same names. For example, say that I want to use the inspect function from both the analysis.utils and the frontend.utils modules. Importing the attributes directly won't work because the second import statement will overwrite the value of inspect in the current scope:

```
# main2.py
from analysis.utils import inspect
from frontend.utils import inspect  # Overwrites!
```

The solution is to use the as clause of the import statement to rename whatever I've imported for the current scope:

```
# main3.py
from analysis.utils import inspect as analysis_inspect
from frontend.utils import inspect as frontend_inspect

value = 33
if analysis_inspect(value) == frontend_inspect(value):
    print('Inspection equal!')
```

The as clause can be used to rename anything retrieved with the import statement, including entire modules. This facilitates accessing namespaced code and makes its identity clear when you use it.

Another approach for avoiding imported name conflicts is to always access names by their highest unique module name. For the example above, this means I'd use basic import statements instead of import from:

```
# main4.py
import analysis.utils
import frontend.utils

value = 33
if (analysis.utils.inspect(value) ==
        frontend.utils.inspect(value)):
    print('Inspection equal!')
```

This approach allows you to avoid the as clause altogether. It also makes it abundantly clear to new readers of the code where each of the similarly named functions is defined.

Stable APIs

The second use of packages in Python is to provide strict, stable APIs for external consumers.

When you're writing an API for wider consumption, such as an open source package (see Item 82: "Know Where to Find Community-Built Modules" for examples), you'll want to provide stable functionality that doesn't change between releases. To ensure that happens, it's important to hide your internal code organization from external users. This way, you can refactor and improve your package's internal modules without breaking existing users.

Python can limit the surface area exposed to API consumers by using the __all__ special attribute of a module or package. The value of __all__ is a list of every name to export from the module as part of its public API. When consuming code executes from foo import *, only the attributes in foo.__all__ will be imported from foo. If __all__ isn't present in foo, then only public attributes—those without a leading underscore—are imported (see Item 42: "Prefer Public Attributes Over Private Ones" for details about that convention).

For example, say that I want to provide a package for calculating collisions between moving projectiles. Here, I define the models module of mypackage to contain the representation of projectiles:

```
# models.py
__all__ = ['Projectile']

class Projectile:
    def __init__(self, mass, velocity):
        self.mass = mass
        self.velocity = velocity
```

I also define a utils module in mypackage to perform operations on the Projectile instances, such as simulating collisions between them:

```
# utils.py
from . models import Projectile

__all__ = ['simulate_collision']

def _dot_product(a, b):
    ...

def simulate_collision(a, b):
    ...
```

Now, I'd like to provide all of the public parts of this API as a set of attributes that are available on the mypackage module. This will allow downstream consumers to always import directly from mypackage instead of importing from mypackage.models or mypackage.utils. This ensures that the API consumer's code will continue to work even if the internal organization of mypackage changes (e.g., models.py is deleted).

To do this with Python packages, you need to modify the __init__.py file in the mypackage directory. This file is what actually becomes the contents of the mypackage module when it's imported. Thus, you can specify an explicit API for mypackage by limiting what you import into __init__.py. Since all of my internal modules already specify __all__, I can expose the public interface of mypackage by simply importing everything from the internal modules and updating __all__ accordingly:

```
# __init__.py
__all__ = []
from . models import *
```

```
__all__ += models.__all__
from . utils import *
__all__ += utils.__all__
```

Here's a consumer of the API that directly imports from `mypackage` instead of accessing the inner modules:

```
# api_consumer.py
from mypackage import *

a = Projectile(1.5, 3)
b = Projectile(4, 1.7)
after_a, after_b = simulate_collision(a, b)
```

Notably, internal-only functions like `mypackage.utils._dot_product` will not be available to the API consumer on `mypackage` because they weren't present in `__all__`. Being omitted from `__all__` also means that they weren't imported by the `from mypackage import *` statement. The internal-only names are effectively hidden.

This whole approach works great when it's important to provide an explicit, stable API. However, if you're building an API for use between your own modules, the functionality of `__all__` is probably unnecessary and should be avoided. The namespacing provided by packages is usually enough for a team of programmers to collaborate on large amounts of code they control while maintaining reasonable interface boundaries.

Beware of `import *`

Import statements like `from x import y` are clear because the source of y is explicitly the x package or module. Wildcard imports like `from foo import *` can also be useful, especially in interactive Python sessions. However, wildcards make code more difficult to understand:

- `from foo import *` hides the source of names from new readers of the code. If a module has multiple `import *` statements, you'll need to check all of the referenced modules to figure out where a name was defined.

- Names from `import *` statements will overwrite any conflicting names within the containing module. This can lead to strange bugs caused by accidental interactions between your code and overlapping names from multiple `import *` statements.

The safest approach is to avoid `import *` in your code and explicitly import names with the `from x import y` style.

Things to Remember

✦ Packages in Python are modules that contain other modules. Packages allow you to organize your code into separate, non-conflicting namespaces with unique absolute module names.

✦ Simple packages are defined by adding an __init__.py file to a directory that contains other source files. These files become the child modules of the directory's package. Package directories may also contain other packages.

✦ You can provide an explicit API for a module by listing its publicly visible names in its __all__ special attribute.

✦ You can hide a package's internal implementation by only importing public names in the package's __init__.py file or by naming internal-only members with a leading underscore.

✦ When collaborating within a single team or on a single codebase, using __all__ for explicit APIs is probably unnecessary.

Item 86: Consider Module-Scoped Code to Configure Deployment Environments

A deployment environment is a configuration in which a program runs. Every program has at least one deployment environment: the *production environment*. The goal of writing a program in the first place is to put it to work in the production environment and achieve some kind of outcome.

Writing or modifying a program requires being able to run it on the computer you use for developing. The configuration of your *development environment* may be very different from that of your production environment. For example, you may be using a tiny single-board computer to develop a program that's meant to run on enormous supercomputers.

Tools like venv (see Item 83: "Use Virtual Environments for Isolated and Reproducible Dependencies") make it easy to ensure that all environments have the same Python packages installed. The trouble is that production environments often require many external assumptions that are hard to reproduce in development environments.

For example, say that I want to run a program in a web server container and give it access to a database. Every time I want to modify my program's code, I need to run a server container, the database schema must be set up properly, and my program needs the password for access. This is a very high cost if all I'm trying to do is verify that a one-line change to my program works correctly.

The best way to work around such issues is to override parts of a program at startup time to provide different functionality depending on the deployment environment. For example, I could have two different __main__ files—one for production and one for development:

```
# dev_main.py
TESTING = True

import db_connection

db = db_connection.Database()
```

```
# prod_main.py
TESTING = False

import db_connection

db = db_connection.Database()
```

The only difference between the two files is the value of the TESTING constant. Other modules in my program can then import the __main__ module and use the value of TESTING to decide how they define their own attributes:

```
# db_connection.py
import __main__

class TestingDatabase:
    ...

class RealDatabase:
    ...

if __main__.TESTING:
    Database = TestingDatabase
else:
    Database = RealDatabase
```

The key behavior to notice here is that code running in module scope—not inside a function or method—is just normal Python code. You can use an if statement at the module level to decide how the module will define names. This makes it easy to tailor modules to your various deployment environments. You can avoid having to reproduce costly assumptions like database configurations when they aren't needed. You can inject local or fake implementations that ease interactive development, or you can use mocks for writing tests (see Item 78: "Use Mocks to Test Code with Complex Dependencies").

Note

When your deployment environment configuration gets really complicated, you should consider moving it out of Python constants (like TESTING) and into dedicated configuration files. Tools like the `configparser` built-in module let you maintain production configurations separately from code, a distinction that's crucial for collaborating with an operations team.

This approach can be used for more than working around external assumptions. For example, if I know that my program must work differently depending on its host platform, I can inspect the sys module before defining top-level constructs in a module:

```python
# db_connection.py
import sys

class Win32Database:
    ...

class PosixDatabase:
    ...

if sys.platform.startswith('win32'):
    Database = Win32Database
else:
    Database = PosixDatabase
```

Similarly, I could use environment variables from `os.environ` to guide my module definitions.

Things to Remember

✦ Programs often need to run in multiple deployment environments that each have unique assumptions and configurations.

✦ You can tailor a module's contents to different deployment environments by using normal Python statements in module scope.

✦ Module contents can be the product of any external condition, including host introspection through the sys and os modules.

Item 87: Define a Root Exception to Insulate Callers from APIs

When you're defining a module's API, the exceptions you raise are just as much a part of your interface as the functions and classes you define (see Item 20: "Prefer Raising Exceptions to Returning None" for an example).

Python has a built-in hierarchy of exceptions for the language and standard library. There's a draw to using the built-in exception types for reporting errors instead of defining your own new types. For example, I could raise a ValueError exception whenever an invalid parameter is passed to a function in one of my modules:

```
# my_module.py
def determine_weight(volume, density):
    if density <= 0:
        raise ValueError('Density must be positive')
    ...
```

In some cases, using ValueError makes sense, but for APIs, it's much more powerful to define a new hierarchy of exceptions. I can do this by providing a root Exception in my module and having all other exceptions raised by that module inherit from the root exception:

```
# my_module.py
class Error(Exception):
    """Base-class for all exceptions raised by this module."""

class InvalidDensityError(Error):
    """There was a problem with a provided density value."""

class InvalidVolumeError(Error):
    """There was a problem with the provided weight value."""

def determine_weight(volume, density):
    if density < 0:
        raise InvalidDensityError('Density must be positive')
    if volume < 0:
        raise InvalidVolumeError('Volume must be positive')
    if volume == 0:
        density / volume
```

Having a root exception in a module makes it easy for consumers of an API to catch all of the exceptions that were raised deliberately. For example, here a consumer of my API makes a function call with a try/except statement that catches my root exception:

```
try:
    weight = my_module.determine_weight(1, -1)
except my_module.Error:
    logging.exception('Unexpected error')

>>>
Unexpected error
```

```
Traceback (most recent call last):
  File ".../example.py", line 3, in <module>
    weight = my_module.determine_weight(1, -1)
  File ".../my_module.py", line 10, in determine_weight
    raise InvalidDensityError('Density must be positive')
InvalidDensityError: Density must be positive
```

Here, the logging.exception function prints the full stack trace of the caught exception so it's easier to debug in this situation. The try/except also prevents my API's exceptions from propagating too far upward and breaking the calling program. It insulates the calling code from my API. This insulation has three helpful effects.

First, root exceptions let callers understand when there's a problem with their usage of an API. If callers are using my API properly, they should catch the various exceptions that I deliberately raise. If they don't handle such an exception, it will propagate all the way up to the insulating except block that catches my module's root exception. That block can bring the exception to the attention of the API consumer, providing an opportunity for them to add proper handling of the missed exception type:

```
try:
    weight = my_module.determine_weight(-1, 1)
except my_module.InvalidDensityError:
    weight = 0
except my_module.Error:
    logging.exception('Bug in the calling code')

>>>
Bug in the calling code
Traceback (most recent call last):
  File ".../example.py", line 3, in <module>
    weight = my_module.determine_weight(-1, 1)
  File ".../my_module.py", line 12, in determine_weight
    raise InvalidVolumeError('Volume must be positive')
InvalidVolumeError: Volume must be positive
```

The second advantage of using root exceptions is that they can help find bugs in an API module's code. If my code only deliberately raises exceptions that I define within my module's hierarchy, then all other types of exceptions raised by my module must be the ones that I didn't intend to raise. These are bugs in my API's code.

Using the try/except statement above will not insulate API consumers from bugs in my API module's code. To do that, the caller needs to add another except block that catches Python's base Exception class.

This allows the API consumer to detect when there's a bug in the API module's implementation that needs to be fixed. The output for this example includes both the logging.exception message and the default interpreter output for the exception since it was re-raised:

```
try:
    weight = my_module.determine_weight(0, 1)
except my_module.InvalidDensityError:
    weight = 0
except my_module.Error:
    logging.exception('Bug in the calling code')
except Exception:
    logging.exception('Bug in the API code!')
    raise  # Re-raise exception to the caller

>>>
Bug in the API code!
Traceback (most recent call last):
  File ".../example.py", line 3, in <module>
    weight = my_module.determine_weight(0, 1)
  File ".../my_module.py", line 14, in determine_weight
    density / volume
ZeroDivisionError: division by zero
Traceback ...
ZeroDivisionError: division by zero
```

The third impact of using root exceptions is future-proofing an API. Over time, I might want to expand my API to provide more specific exceptions in certain situations. For example, I could add an Exception subclass that indicates the error condition of supplying negative densities:

```
# my_module.py
...

class NegativeDensityError(InvalidDensityError):
    """A provided density value was negative."""


...


def determine_weight(volume, density):
    if density < 0:
        raise NegativeDensityError('Density must be positive')
    ...
```

The calling code will continue to work exactly as before because it already catches InvalidDensityError exceptions (the parent class of NegativeDensityError). In the future, the caller could decide to special-case the new type of exception and change the handling behavior accordingly:

```
try:
    weight = my_module.determine_weight(1, -1)
except my_module.NegativeDensityError:
    raise ValueError('Must supply non-negative density')
except my_module.InvalidDensityError:
    weight = 0
except my_module.Error:
    logging.exception('Bug in the calling code')
except Exception:
    logging.exception('Bug in the API code!')
    raise

>>>
Traceback ...
NegativeDensityError: Density must be positive

The above exception was the direct cause of the following
➥exception:

Traceback ...
ValueError: Must supply non-negative density
```

I can take API future-proofing further by providing a broader set of exceptions directly below the root exception. For example, imagine that I have one set of errors related to calculating weights, another related to calculating volume, and a third related to calculating density:

```
# my_module.py
class Error(Exception):
    """Base-class for all exceptions raised by this module."""

class WeightError(Error):
    """Base-class for weight calculation errors."""

class VolumeError(Error):
    """Base-class for volume calculation errors."""

class DensityError(Error):
    """Base-class for density calculation errors."""

...
```

Specific exceptions would inherit from these general exceptions. Each intermediate exception acts as its own kind of root exception. This makes it easier to insulate layers of calling code from API code based on broad functionality. This is much better than having all callers catch a long list of very specific Exception subclasses.

Things to Remember

✦ Defining root exceptions for modules allows API consumers to insulate themselves from an API.

✦ Catching root exceptions can help you find bugs in code that consumes an API.

✦ Catching the Python Exception base class can help you find bugs in API implementations.

✦ Intermediate root exceptions let you add more specific types of exceptions in the future without breaking your API consumers.

Item 88: Know How to Break Circular Dependencies

Inevitably, while you're collaborating with others, you'll find a mutual interdependence between modules. It can even happen while you work by yourself on the various parts of a single program.

For example, say that I want my GUI application to show a dialog box for choosing where to save a document. The data displayed by the dialog could be specified through arguments to my event handlers. But the dialog also needs to read global state, such as user preferences, to know how to render properly.

Here, I define a dialog that retrieves the default document save location from global preferences:

```
# dialog.py
import app

class Dialog:
    def __init__(self, save_dir):
        self.save_dir = save_dir
    ...

save_dialog = Dialog(app.prefs.get('save_dir'))

def show():
    ...
```

The problem is that the app module that contains the prefs object also imports the dialog class in order to show the same dialog on program start:

```
# app.py
import dialog

class Prefs:
    ...
    def get(self, name):
        ...


prefs = Prefs()
dialog.show()
```

It's a circular dependency. If I try to import the app module from my main program like this:

```
# main.py
import app
```

I get an exception:

```
>>>
$ python3 main.py
Traceback (most recent call last):
  File ".../main.py", line 17, in <module>
    import app
  File ".../app.py", line 17, in <module>
    import dialog
  File ".../dialog.py", line 23, in <module>
    save_dialog = Dialog(app.prefs.get('save_dir'))
AttributeError: partially initialized module 'app' has no
➥attribute 'prefs' (most likely due to a circular import)
```

To understand what's happening here, you need to know how Python's import machinery works in general (see the importlib built-in package for the full details). When a module is imported, here's what Python actually does, in depth-first order:

1. Searches for a module in locations from sys.path

2. Loads the code from the module and ensures that it compiles

3. Creates a corresponding empty module object

4. Inserts the module into sys.modules

5. Runs the code in the module object to define its contents

The problem with a circular dependency is that the attributes of a module aren't defined until the code for those attributes has executed (after step 5). But the module can be loaded with the import statement immediately after it's inserted into sys.modules (after step 4).

In the example above, the app module imports dialog before defining anything. Then, the dialog module imports app. Since app still hasn't finished running—it's currently importing dialog—the app module is empty (from step 4). The AttributeError is raised (during step 5 for dialog) because the code that defines prefs hasn't run yet (step 5 for app isn't complete).

The best solution to this problem is to refactor the code so that the prefs data structure is at the bottom of the dependency tree. Then, both app and dialog can import the same utility module and avoid any circular dependencies. But such a clear division isn't always possible or could require too much refactoring to be worth the effort.

There are three other ways to break circular dependencies.

Reordering Imports

The first approach is to change the order of imports. For example, if I import the dialog module toward the bottom of the app module, after the app module's other contents have run, the AttributeError goes away:

```
# app.py
class Prefs:
    ...

prefs = Prefs()

import dialog  # Moved
dialog.show()
```

This works because, when the dialog module is loaded late, its recursive import of app finds that app.prefs has already been defined (step 5 is mostly done for app).

Although this avoids the AttributeError, it goes against the PEP 8 style guide (see Item 2: "Follow the PEP 8 Style Guide"). The style guide suggests that you always put imports at the top of your Python files. This makes your module's dependencies clear to new readers of the code. It also ensures that any module you depend on is in scope and available to all the code in your module.

Having imports later in a file can be brittle and can cause small changes in the ordering of your code to break the module entirely. I suggest not using import reordering to solve your circular dependency issues.

Import, Configure, Run

A second solution to the circular imports problem is to have modules minimize side effects at import time. I can have my modules only define functions, classes, and constants. I avoid actually running any functions at import time. Then, I have each module provide a configure function that I call once all other modules have finished importing. The purpose of configure is to prepare each module's state by accessing the attributes of other modules. I run configure after all modules have been imported (step 5 is complete), so all attributes must be defined.

Here, I redefine the dialog module to only access the prefs object when configure is called:

```
# dialog.py
import app

class Dialog:
    ...

save_dialog = Dialog()

def show():
    ...

def configure():
    save_dialog.save_dir = app.prefs.get('save_dir')
```

I also redefine the app module to not run activities on import:

```
# app.py
import dialog

class Prefs:
    ...

prefs = Prefs()

def configure():
    ...
```

Finally, the main module has three distinct phases of execution— import everything, configure everything, and run the first activity:

```
# main.py
import app
import dialog

app.configure()
dialog.configure()

dialog.show()
```

This works well in many situations and enables patterns like *dependency injection*. But sometimes it can be difficult to structure your code so that an explicit configure step is possible. Having two distinct phases within a module can also make your code harder to read because it separates the definition of objects from their configuration.

Dynamic Import

The third—and often simplest—solution to the circular imports problem is to use an import statement within a function or method. This is called a *dynamic import* because the module import happens while the program is running, not while the program is first starting up and initializing its modules.

Here, I redefine the dialog module to use a dynamic import. The dialog.show function imports the app module at runtime instead of the dialog module importing app at initialization time:

```
# dialog.py
class Dialog:
    ...

save_dialog = Dialog()

def show():
    import app  # Dynamic import
    save_dialog.save_dir = app.prefs.get('save_dir')
    ...
```

The app module can now be the same as it was in the original example. It imports dialog at the top and calls dialog.show at the bottom:

```
# app.py
import dialog
```

```
class Prefs:
    ...

prefs = Prefs()
dialog.show()
```

This approach has a similar effect to the import, configure, and run steps from before. The difference is that it requires no structural changes to the way the modules are defined and imported. I'm simply delaying the circular import until the moment I must access the other module. At that point, I can be pretty sure that all other modules have already been initialized (step 5 is complete for everything).

In general, it's good to avoid dynamic imports like this. The cost of the import statement is not negligible and can be especially bad in tight loops. By delaying execution, dynamic imports also set you up for surprising failures at runtime, such as SyntaxError exceptions long after your program has started running (see Item 76: "Verify Related Behaviors in TestCase Subclasses" for how to avoid that). However, these downsides are often better than the alternative of restructuring your entire program.

Things to Remember

✦ Circular dependencies happen when two modules must call into each other at import time. They can cause your program to crash at startup.

✦ The best way to break a circular dependency is by refactoring mutual dependencies into a separate module at the bottom of the dependency tree.

✦ Dynamic imports are the simplest solution for breaking a circular dependency between modules while minimizing refactoring and complexity.

Item 89: Consider warnings to Refactor and Migrate Usage

It's natural for APIs to change in order to satisfy new requirements that meet formerly unanticipated needs. When an API is small and has few upstream or downstream dependencies, making such changes is straightforward. One programmer can often update a small API and all of its callers in a single commit.

However, as a codebase grows, the number of callers of an API can be so large or fragmented across source repositories that it's infeasible or impractical to make API changes in lockstep with updating callers to match. Instead, you need a way to notify and encourage the people that you collaborate with to refactor their code and migrate their API usage to the latest forms.

For example, say that I want to provide a module for calculating how far a car will travel at a given average speed and duration. Here, I define such a function and assume that speed is in miles per hour and duration is in hours:

```
def print_distance(speed, duration):
    distance = speed * duration
    print(f'{distance} miles')

print_distance(5, 2.5)

>>>
12.5 miles
```

Imagine that this works so well that I quickly gather a large number of dependencies on this function. Other programmers that I collaborate with need to calculate and print distances like this all across our shared codebase.

Despite its success, this implementation is error prone because the units for the arguments are implicit. For example, if I wanted to see how far a bullet travels in 3 seconds at 1000 meters per second, I would get the wrong result:

```
print_distance(1000, 3)

>>>
3000 miles
```

I can address this problem by expanding the API of print_distance to include optional keyword arguments (see Item 23: "Provide Optional Behavior with Keyword Arguments" and Item 25: "Enforce Clarity with Keyword-Only and Positional-Only Arguments") for the units of speed, duration, and the computed distance to print out:

```
CONVERSIONS = {
    'mph': 1.60934 / 3600 * 1000,   # m/s
    'hours': 3600,                  # seconds
    'miles': 1.60934 * 1000,        # m
    'meters': 1,                    # m
    'm/s': 1,                       # m
    'seconds': 1,                   # s
}
```

```
def convert(value, units):
    rate = CONVERSIONS[units]
    return rate * value

def localize(value, units):
    rate = CONVERSIONS[units]
    return value / rate

def print_distance(speed, duration, *,
                   speed_units='mph',
                   time_units='hours',
                   distance_units='miles'):
    norm_speed = convert(speed, speed_units)
    norm_duration = convert(duration, time_units)
    norm_distance = norm_speed * norm_duration
    distance = localize(norm_distance, distance_units)
    print(f'{distance} {distance_units}')
```

Now, I can modify the speeding bullet call to produce an accurate result with a unit conversion to miles:

```
print_distance(1000, 3,
               speed_units='meters',
               time_units='seconds')
```

```
>>>
1.8641182099494205 miles
```

It seems like requiring units to be specified for this function is a much better way to go. Making them explicit reduces the likelihood of errors and is easier for new readers of the code to understand. But how can I migrate all callers of the API over to always specifying units? How do I minimize breakage of any code that's dependent on print_distance while also encouraging callers to adopt the new units arguments as soon as possible?

For this purpose, Python provides the built-in warnings module. Using warnings is a programmatic way to inform other programmers that their code needs to be modified due to a change to an underlying library that they depend on. While exceptions are primarily for automated error handling by machines (see Item 87: "Define a Root Exception to Insulate Callers from APIs"), warnings are all about communication between humans about what to expect in their collaboration with each other.

I can modify print_distance to issue warnings when the optional keyword arguments for specifying units are not supplied. This way, the arguments can continue being optional temporarily (see Item 24: "Use None and Docstrings to Specify Dynamic Default Arguments" for background), while providing an explicit notice to people running dependent programs that they should expect breakage in the future if they fail to take action:

```
import warnings

def print_distance(speed, duration, *,
                   speed_units=None,
                   time_units=None,
                   distance_units=None):
    if speed_units is None:
        warnings.warn(
            'speed_units required', DeprecationWarning)
        speed_units = 'mph'

    if time_units is None:
        warnings.warn(
            'time_units required', DeprecationWarning)
        time_units = 'hours'

    if distance_units is None:
        warnings.warn(
            'distance_units required', DeprecationWarning)
        distance_units = 'miles'

    norm_speed = convert(speed, speed_units)
    norm_duration = convert(duration, time_units)
    norm_distance = norm_speed * norm_duration
    distance = localize(norm_distance, distance_units)
    print(f'{distance} {distance_units}')
```

I can verify that this code issues a warning by calling the function with the same arguments as before and capturing the sys.stderr output from the warnings module:

```
import contextlib
import io

fake_stderr = io.StringIO()
with contextlib.redirect_stderr(fake_stderr):
```

```
    print_distance(1000, 3,
                   speed_units='meters',
                   time_units='seconds')

print(fake_stderr.getvalue())

>>>
1.8641182099494205 miles
.../example.py:97: DeprecationWarning: distance_units required
  warnings.warn(
```

Adding warnings to this function required quite a lot of repetitive boilerplate that's hard to read and maintain. Also, the warning message indicates the line where warning.warn was called, but what I really want to point out is where the call to print_distance was made *without* soon-to-be-required keyword arguments.

Luckily, the warnings.warn function supports the stacklevel parameter, which makes it possible to report the correct place in the stack as the cause of the warning. stacklevel also makes it easy to write functions that can issue warnings on behalf of other code, reducing boilerplate. Here, I define a helper function that warns if an optional argument wasn't supplied and then provides a default value for it:

```
def require(name, value, default):
    if value is not None:
        return value
    warnings.warn(
        f'{name} will be required soon, update your code',
        DeprecationWarning,
        stacklevel=3)
    return default

def print_distance(speed, duration, *,
                   speed_units=None,
                   time_units=None,
                   distance_units=None):
    speed_units = require('speed_units', speed_units, 'mph')
    time_units = require('time_units', time_units, 'hours')
    distance_units = require(
        'distance_units', distance_units, 'miles')

    norm_speed = convert(speed, speed_units)
    norm_duration = convert(duration, time_units)
    norm_distance = norm_speed * norm_duration
    distance = localize(norm_distance, distance_units)
    print(f'{distance} {distance_units}')
```

I can verify that this propagates the proper offending line by inspecting the captured output:

```
import contextlib
import io

fake_stderr = io.StringIO()
with contextlib.redirect_stderr(fake_stderr):
    print_distance(1000, 3,
                   speed_units='meters',
                   time_units='seconds')

print(fake_stderr.getvalue())

>>>
1.8641182099494205 miles
.../example.py:174: DeprecationWarning: distance_units will be
➥required soon, update your code
  print_distance(1000, 3,
```

The warnings module also lets me configure what should happen when a warning is encountered. One option is to make all warnings become errors, which raises the warning as an exception instead of printing it out to sys.stderr:

```
warnings.simplefilter('error')
try:
    warnings.warn('This usage is deprecated',
                  DeprecationWarning)
except DeprecationWarning:
    pass  # Expected
```

This exception-raising behavior is especially useful for automated tests in order to detect changes in upstream dependencies and fail tests accordingly. Using such test failures is a great way to make it clear to the people you collaborate with that they will need to update their code. You can use the -W error command-line argument to the Python interpreter or the PYTHONWARNINGS environment variable to apply this policy:

```
$ python -W error example_test.py
Traceback (most recent call last):
  File ".../example_test.py", line 6, in <module>
    warnings.warn('This might raise an exception!')
UserWarning: This might raise an exception!
```

Once the people responsible for code that depends on a deprecated API are aware that they'll need to do a migration, they can tell the warnings module to ignore the error by using the `simplefilter` and `filterwarnings` functions:

```
warnings.simplefilter('ignore')
warnings.warn('This will not be printed to stderr')
```

After a program is deployed into production, it doesn't make sense for warnings to cause errors because they might crash the program at a critical time. Instead, a better approach is to replicate warnings into the `logging` built-in module. Here, I accomplish this by calling the `logging.captureWarnings` function and configuring the corresponding `'py.warnings'` logger:

```
import logging


fake_stderr = io.StringIO()
handler = logging.StreamHandler(fake_stderr)
formatter = logging.Formatter(
    '%(asctime)-15s WARNING] %(message)s')
handler.setFormatter(formatter)

logging.captureWarnings(True)
logger = logging.getLogger('py.warnings')
logger.addHandler(handler)
logger.setLevel(logging.DEBUG)

warnings.resetwarnings()
warnings.simplefilter('default')
warnings.warn('This will go to the logs output')

print(fake_stderr.getvalue())

>>>
2019-06-11 19:48:19,132 WARNING] .../example.py:227:
➥UserWarning: This will go to the logs output
  warnings.warn('This will go to the logs output')
```

Using logging to capture warnings ensures that any error reporting systems that my program already has in place will also receive notice of important warnings in production. This can be especially useful if my tests don't cover every edge case that I might see when the program is undergoing real usage.

API library maintainers should also write unit tests to verify that warnings are generated under the correct circumstances with clear and actionable messages (see Item 76: "Verify Related Behaviors in TestCase Subclasses"). Here, I use the `warnings.catch_warnings` function as a context manager (see Item 66: "Consider `contextlib` and `with` Statements for Reusable `try/finally` Behavior" for background) to wrap a call to the `require` function that I defined above:

```
with warnings.catch_warnings(record=True) as found_warnings:
    found = require('my_arg', None, 'fake units')
    expected = 'fake units'
    assert found == expected
```

Once I've collected the warning messages, I can verify that their number, detail messages, and categories match my expectations:

```
assert len(found_warnings) == 1
single_warning = found_warnings[0]
assert str(single_warning.message) == (
    'my_arg will be required soon, update your code')
assert single_warning.category == DeprecationWarning
```

Things to Remember

✦ The `warnings` module can be used to notify callers of your API about deprecated usage. Warning messages encourage such callers to fix their code before later changes break their programs.

✦ Raise warnings as errors by using the `-W error` command-line argument to the Python interpreter. This is especially useful in automated tests to catch potential regressions of dependencies.

✦ In production, you can replicate warnings into the `logging` module to ensure that your existing error reporting systems will capture warnings at runtime.

✦ It's useful to write tests for the warnings that your code generates to make sure that they'll be triggered at the right time in any of your downstream dependencies.

Item 90: Consider Static Analysis via typing to Obviate Bugs

Providing documentation is a great way to help users of an API understand how to use it properly (see Item 84: "Write Docstrings for Every Function, Class, and Module"), but often it's not enough, and incorrect usage still causes bugs. Ideally, there would be a programmatic mechanism to verify that callers are using your APIs the right way, and that you are using your downstream dependencies correctly. Many programming languages address part of this need with compile-time type checking, which can identify and eliminate some categories of bugs.

Historically Python has focused on dynamic features and has not provided compile-time type safety of any kind. However, more recently Python has introduced special syntax and the built-in `typing` module, which allow you to annotate variables, class fields, functions, and methods with type information. These *type hints* allow for *gradual typing*, where a codebase can be incrementally updated to specify types as desired.

The benefit of adding type information to a Python program is that you can run *static analysis* tools to ingest a program's source code and identify where bugs are most likely to occur. The `typing` built-in module doesn't actually implement any of the type checking functionality itself. It merely provides a common library for defining types, including generics, that can be applied to Python code and consumed by separate tools.

Much as there are multiple distinct implementations of the Python interpreter (e.g., CPython, PyPy), there are multiple implementa-tions of static analysis tools for Python that use `typing`. As of the time of this writing, the most popular tools are `mypy`, `pytype`, `pyright`, and `pyre`. For the typing examples in this book, I've used `mypy` with the `--strict` flag, which enables all of the various warnings supported by the tool. Here's an example of what running the command line looks like:

```
$ python3 -m mypy --strict example.py
```

These tools can be used to detect a large number of common errors before a program is ever run, which can provide an added layer of safety in addition to having good unit tests (see Item 76: "Verify Related Behaviors in `TestCase` Subclasses"). For example, can you find the bug in this simple function that causes it to compile fine but throw an exception at runtime?

```
def subtract(a, b):
    return a - b


subtract(10, '5')

>>>
Traceback ...
TypeError: unsupported operand type(s) for -: 'int' and 'str'
```

Parameter and variable type annotations are delineated with a colon (such as name: type). Return value types are specified with -> type following the argument list. Using such type annotations and mypy, I can easily spot the bug:

```
def subtract(a: int, b: int) -> int:  # Function annotation
    return a - b
```

```
subtract(10, '5')  # Oops: passed string value
```

```
$ python3 -m mypy --strict example.py
.../example.py:4: error: Argument 2 to "subtract" has
incompatible type "str"; expected "int"
```

Another common mistake, especially for programmers who have recently moved from Python 2 to Python 3, is mixing bytes and str instances together (see Item 3: "Know the Differences Between bytes and str"). Do you see the problem in this example that causes a run-time error?

```
def concat(a, b):
    return a + b
```

```
concat('first', b'second')
```

```
>>>
Traceback ...
TypeError: can only concatenate str (not "bytes") to str
```

Using type hints and mypy, this issue can be detected statically before the program runs:

```
def concat(a: str, b: str) -> str:
    return a + b
```

```
concat('first', b'second')  # Oops: passed bytes value
```

```
$ python3 -m mypy --strict example.py
.../example.py:4: error: Argument 2 to "concat" has
➥incompatible type "bytes"; expected "str"
```

Type annotations can also be applied to classes. For example, this class has two bugs in it that will raise exceptions when the program is run:

```
class Counter:
    def __init__(self):
        self.value = 0
```

```
    def add(self, offset):
        value += offset

    def get(self) -> int:
        self.value
```

The first one happens when I call the add method:

```
counter = Counter()
counter.add(5)
```

```
>>>
Traceback ...
UnboundLocalError: local variable 'value' referenced before
➥assignment
```

The second bug happens when I call get:

```
counter = Counter()
found = counter.get()
assert found == 0, found
```

```
>>>
Traceback ...
AssertionError: None
```

Both of these problems are easily found by mypy:

```
class Counter:
    def __init__(self) -> None:
        self.value: int = 0  # Field / variable annotation

    def add(self, offset: int) -> None:
        value += offset       # Oops: forgot "self."

    def get(self) -> int:
        self.value            # Oops: forgot "return"

counter = Counter()
counter.add(5)
counter.add(3)
assert counter.get() == 8
```

```
$ python3 -m mypy --strict example.py
.../example.py:6: error: Name 'value' is not defined
.../example.py:8: error: Missing return statement
```

One of the strengths of Python's dynamism is the ability to write generic functionality that operates on duck types (see Item 15: "Be Cautious When Relying on dict Insertion Ordering" and Item 43: "Inherit from collections.abc for Custom Container Types"). This allows one implementation to accept a wide range of types, saving a lot of duplicative effort and simplifying testing. Here, I've defined such a generic function for combining values from a list. Do you understand why the last assertion fails?

```python
def combine(func, values):
    assert len(values) > 0

    result = values[0]
    for next_value in values[1:]:
        result = func(result, next_value)

    return result

def add(x, y):
    return x + y

inputs = [1, 2, 3, 4j]
result = combine(add, inputs)
assert result == 10, result  # Fails

>>>
Traceback ...
AssertionError: (6+4j)
```

I can use the typing module's support for generics to annotate this function and detect the problem statically:

```python
from typing import Callable, List, TypeVar

Value = TypeVar('Value')
Func = Callable[[Value, Value], Value]

def combine(func: Func[Value], values: List[Value]) -> Value:
    assert len(values) > 0

    result = values[0]
    for next_value in values[1:]:
        result = func(result, next_value)

    return result
```

```
Real = TypeVar('Real', int, float)

def add(x: Real, y: Real) -> Real:
    return x + y

inputs = [1, 2, 3, 4j]  # Oops: included a complex number
result = combine(add, inputs)
assert result == 10
$ python3 -m mypy --strict example.py
.../example.py:21: error: Argument 1 to "combine" has
➥incompatible type "Callable[[Real, Real], Real]"; expected
➥"Callable[[complex, complex], complex]"
```

Another extremely common error is to encounter a None value when you thought you'd have a valid object (see Item 20: "Prefer Raising Exceptions to Returning None"). This problem can affect seemingly simple code. Do you see the issue here?

```
def get_or_default(value, default):
    if value is not None:
        return value
    return value

found = get_or_default(3, 5)
assert found == 3

found = get_or_default(None, 5)
assert found == 5, found  # Fails

>>>
Traceback ...
AssertionError: None
```

The typing module supports *option types*, which ensure that programs only interact with values after proper null checks have been performed. This allows mypy to infer that there's a bug in this code: The type used in the return statement must be None, and that doesn't match the int type required by the function signature:

```
from typing import Optional

def get_or_default(value: Optional[int],
                   default: int) -> int:
    if value is not None:
        return value
    return value  # Oops: should have returned "default"
```

```
$ python3 -m mypy --strict example.py
.../example.py:7: error: Incompatible return value type (got
➡"None", expected "int")
```

A wide variety of other options are available in the typing module. Notably, exceptions are not included. Unlike Java, which has checked exceptions that are enforced at the API boundary of every method, Python's type annotations are more similar to C#'s: Exceptions are not considered part of an interface's definition. Thus, if you want to verify that you're raising and catching exceptions properly, you need to write tests.

One common gotcha in using the typing module occurs when you need to deal with forward references (see Item 88: "Know How to Break Circular Dependencies" for a similar problem). For example, imagine that I have two classes and one holds a reference to the other:

```
class FirstClass:
    def __init__(self, value):
        self.value = value

class SecondClass:
    def __init__(self, value):
        self.value = value

second = SecondClass(5)
first = FirstClass(second)
```

If I apply type hints to this program and run mypy it will say that there are no issues:

```
class FirstClass:
    def __init__(self, value: SecondClass) -> None:
        self.value = value

class SecondClass:
    def __init__(self, value: int) -> None:
        self.value = value

second = SecondClass(5)
first = FirstClass(second)

$ python3 -m mypy --strict example.py
```

However, if you actually try to run this code, it will fail because SecondClass is referenced by the type annotation in the FirstClass.__init__ method's parameters before it's actually defined:

```
class FirstClass:
    def __init__(self, value: SecondClass) -> None:  # Breaks
        self.value = value


class SecondClass:
    def __init__(self, value: int) -> None:
        self.value = value


second = SecondClass(5)
first = FirstClass(second)

>>>
Traceback ...
NameError: name 'SecondClass' is not defined
```

One workaround supported by these static analysis tools is to use a string as the type annotation that contains the forward reference. The string value is later parsed and evaluated to extract the type information to check:

```
class FirstClass:
    def __init__(self, value: 'SecondClass') -> None:  # OK
        self.value = value


class SecondClass:
    def __init__(self, value: int) -> None:
        self.value = value


second = SecondClass(5)
first = FirstClass(second)
```

A better approach is to use from __future__ import annotations, which is available in Python 3.7 and will become the default in Python 4. This instructs the Python interpreter to completely ignore the values supplied in type annotations when the program is being run. This resolves the forward reference problem and provides a performance improvement at program start time:

```
from __future__ import annotations

class FirstClass:
    def __init__(self, value: SecondClass) -> None:  # OK
        self.value = value
```

```
class SecondClass:
    def __init__(self, value: int) -> None:
        self.value = value

second = SecondClass(5)
first = FirstClass(second)
```

Now that you've seen how to use type hints and their potential benefits, it's important to be thoughtful about when to use them. Here are some of the best practices to keep in mind:

- It's going to slow you down if you try to use type annotations from the start when writing a new piece of code. A general strategy is to write a first version without annotations, then write tests, and then add type information where it's most valuable.

- Type hints are most important at the boundaries of a codebase, such as an API you provide that many callers (and thus other people) depend on. Type hints complement integration tests (see Item 77: "Isolate Tests from Each Other with setUp, tearDown, setUpModule, and tearDownModule") and warnings (see Item 89: "Consider warnings to Refactor and Migrate Usage") to ensure that your API callers aren't surprised or broken by your changes.

- It can be useful to apply type hints to the most complex and error-prone parts of your codebase that aren't part of an API. However, it may not be worth striving for 100% coverage in your type annotations because you'll quickly encounter diminishing returns.

- If possible, you should include static analysis as part of your automated build and test system to ensure that every commit to your codebase is vetted for errors. In addition, the configuration used for type checking should be maintained in the repository to ensure that all of the people you collaborate with are using the same rules.

- As you add type information to your code, it's important to run the type checker as you go. Otherwise, you may nearly finish sprinkling type hints everywhere and then be hit by a huge wall of errors from the type checking tool, which can be disheartening and make you want to abandon type hints altogether.

Finally, it's important to acknowledge that in many situations, you may not need or want to use type annotations at all. For small programs, ad-hoc code, legacy codebases, and prototypes, type hints may require far more effort than they're worth.

Things to Remember

✦ Python has special syntax and the typing built-in module for annotating variables, fields, functions, and methods with type information.

✦ Static type checkers can leverage type information to help you avoid many common bugs that would otherwise happen at runtime.

✦ There are a variety of best practices for adopting types in your programs, using them in APIs, and making sure they don't get in the way of your productivity.